Birds of
OMAN

Dedicated to

*BirdLife International's
Middle East Conservation Programme*

HELM FIELD GUIDES

Birds of
OMAN

Jens Eriksen and Richard Porter

with acknowledgement to
Simon Aspinall

Illustrated by
John Gale, Mike Langman and Brian Small

HELM
LONDON · OXFORD · NEW YORK · NEW DELHI · SYDNEY

HELM
Bloomsbury Publishing Plc
50 Bedford Square, London, WC1B 3DP, UK
Bloomsbury Publishing Ireland Limited,
29 Earlsfort Terrace, Dublin 2, D02 AY28, Ireland

BLOOMSBURY, HELM and the Helm logo are trademarks of Bloomsbury Publishing Plc

First published in the United Kingdom 2017

Copyright © Jens Eriksen and Richard Porter, 2017
Illustrations © John Gale, Mike Langman and Brian Small, 2017

Jens Eriksen and Richard Porter have asserted their right under the
Copyright, Designs and Patents Act, 1988, to be identified as Authors of this work

All rights reserved. No part of this publication may be: i) reproduced or transmitted in any form, electronic or mechanical, including photocopying, recording or by means of any information storage or retrieval system without prior permission in writing from the publishers; or ii) used or reproduced in any way for the training, development or operation of artificial intelligence (AI) technologies, including generative AI technologies. The rights holders expressly reserve this publication from the text and data mining exception as per Article 4(3) of the Digital Single Market Directive (EU) 2019/790

Bloomsbury Publishing Plc does not have any control over, or responsibility for, any third-party websites referred to or in this book. All internet addresses given in this book were correct at the time of going to press. The authors and publisher regret any inconvenience caused if addresses have changed or sites have ceased to exist, but can accept no responsibility for any such changes

A catalogue record for this book is available from the British Library

ISBN: PB: 978-1-4729-3753-7;
ePDF: 978-1-4729-3754-4; ePub: 978-1-4729-8321-3

4 6 8 10 9 7 5 3

Designed by Julie Dando, Fluke Art
Printed and bound in India by Thomson Press Ltd.

Cover artwork
Front: Arabian Golden-winged Grosbeak (John Gale)
Back, top to bottom: Hume's Wheatear (Brian Small); Crowned Sandgrouse (John Gale); Omani Owl (Brian Small); Sooty Falcon (John Gale)

To find out more about our authors and books visit www.bloomsbury.com and sign up for our newsletters
For product safety related questions contact productsafety@bloomsbury.com

CONTENTS

	Plate	Page
FOREWORD		7
ACKNOWLEDGEMENTS		8
INTRODUCTION		9
Taxonomy, nomenclature and sequence		11
Illustrations and identification text		11
Bird topography		12
Voice		13
Habitat		13
Maps and status		14
Code of conduct and submitting records		15
SPECIES ACCOUNTS		16
Ducks, geese and swans ANATIDAE	1–5	16–24
Buttonquails TURNICIDAE	5	24
Partridges, francolins and quails PHASIANIDAE	6	26
Storm-petrels HYDROBATIDAE	7	28
Petrels and shearwaters PROCELLARIIDAE	8–9	30–32
Grebes PODICIPEDIDAE	10	34
Flamingos PHOENICOPTERIDAE	10	34
Storks CICONIIDAE	11	36
Cranes GRUIDAE	11	36
Ibises and spoonbills THRESKIORNITHIDAE	12	38
Herons, egrets and bitterns ARDEIDAE	13–16	40–46
Pelicans PELECANIDAE	17	48
Frigatebirds FREGATIDAE	17	48
Gannets and boobies SULIDAE	18	50
Cormorants PHALACROCORACIDAE	18	50
Vultures, hawks, buzzards and eagles ACCIPITRIDAE	19–27	52–68
Falcons FALCONIDAE	28–30	70–74
Bustards OTIDIDAE	31	76
Stone-curlews and thick-knees BURHINIDAE	31	76
Rails, crakes, moorhens and coots RALLIDAE	32–34	78–82
Oystercatchers HAEMATOPODIDAE	35	84
Crab-plover DROMADIDAE	35	84
Stilts and avocets RECURVIROSTRIDAE	35	84
Plovers CHARADRIIDAE	36–39	84–92
Jacanas JACANIDAE	37	86
Snipes and sandpipers SCOLOPACIDAE	40–48	94–110
Coursers and pratincoles GLAREOLIDAE	49	112
Gulls, terns and skimmers LARIDAE	50–58	114–130
Skuas STERCORARIIDAE	59	132
Sandgrouse PTEROCLIDAE	60	134
Pigeons and doves COLUMBIDAE	61–63	136–140

Parrots PSITTACIDAE	63	140
Cuckoos CUCULIDAE	64–65	142–144
Nightjars CAPRIMULGIDAE	65	144
Barn owls TYTONIDAE	66	146
Owls STRIGIDAE	66–68	146–150
Swifts APODIDAE	69	152
Rollers CORACIIDAE	70	154
Hoopoes UPUPIDAE	70	154
Woodpeckers PICIDAE	70	154
Kingfishers ALCEDINIDAE	71	156
Bee-eaters MEROPIDAE	72	158
Bush-shrikes MALACONOTIDAE	72	158
Shrikes LANIIDAE	73–74	160–162
Orioles ORIOLIDAE	75	164
Drongos DICRURIDAE	75	164
Laughingthrushes and babblers LEIOTHRICHIDAE	75	164
Monarchs and paradise-flycatchers MONARCHIDAE	76	166
Hypocolius HYPOCOLIIDAE	76	166
Penduline-tits REMIZIDAE	76	166
Bulbuls PYCNONOTIDAE	76	166
Crows and jays CORVIDAE	77	168
Larks ALAUDIDAE	78–80	170–174
Swallows and martins HIRUNDINIDAE	81–82	176–178
Bush warblers CETTIIDAE	83	180
Streaked Scrub Warbler SCOTOCERCIDAE	83	180
Cisticolas CISTICOLIDAE	83	180
Leaf warblers PHYLLOSCOPIDAE	84–85	182–184
Reed warblers ACROCEPHALIDAE	86–89	186–192
Grassbirds and allies LOCUSTELLIDAE	87	188
Sylviid babblers SYLVIIDAE	89–91	192–196
White-eyes ZOSTEROPIDAE	91	196
Starlings STURNIDAE	92–93	198–200
Thrushes TURDIDAE	94–95	202–204
Chats and flycatchers MUSCICAPIDAE	95–104	204–222
Sunbirds NECTARINIIDAE	104	222
Sparrows PASSERIDAE	105	224
Weavers PLOCEIDAE	106	226
Waxbills, munias and allies ESTRILDIDAE	107	228
Accentors PRUNELLIDAE	108	230
Wagtails and pipits MOTACILLIDAE	109–112	232–238
Finches FRINGILLIDAE	113–114	240–242
Buntings EMBERIZIDAE	114–116	242–246
REFERENCES AND FURTHER READING		248
CHECKLIST OF THE BIRDS OF OMAN		251
INDEX		262

FOREWORD

It is a great honour for me to write the foreword for a book that I know will bring pleasure to many people. Wildlife field guides are one of the building blocks of conservation. If we don't know what we have, we cannot easily conserve it and as a passionate wildlife conservationist any book that increases our knowledge of the wonders of our natural world must be applauded.

In a region that has faced so much conflict and sadness in recent years, books that help with an appreciation of our natural heritage and wonders of our world are to be treasured.

My hope is that the book will be translated into Arabic – as was recently the case with *Birds of the Middle East*, so that the knowledge it contains will be available to all our Arab friends.

Oman is a wonderful country and so rich in bird biodiversity. I hope this book inspires its resident people to love and respect their birds, decision-makers to protect the places where they live and visitors to take away treasured memories.

Khaled Anis Irani
Chairman
BirdLife International

Hume's Wheatear, Al-Hajar mountains, January (Rebecca Reeves & Jim Martin)

ACKNOWLEDGEMENTS

In preparing this field guide we have received help and encouragement from many people. We would particularly like to thank the following who made invaluable contributions and provided records, some over many years, (and sincere apologies to anyone inadvertently omitted): Zahran Al-Abdulasalam, Mundher Al-Abri, Waheed Al-Fazari, Simon Aspinall, John Atkins, Malal Al-Kindi, Anders Blomdahl, Ian Brown, John Bryan, Oscar Campbell, Peter Castell, Bill Clark, Elaine and Peter Cowan, Matt Cummins, Ralph Daly, David Diskin, Tobias Epple, Wouter Faveyts, Rob Felix, Dick Forsman, David Foster, Hilary Fry, Michael Gallagher, Con Greaves, Andrew Grieve, Chris Griffiths, Franziska Güpner, Dominic Harmer, Ian Harrison, Derek Harvey, Erik Hirschfeld, Mike Jennings, Rolf Jensen, Andrew Lassey, Cecile Lazaro, Donnie Mackenzie, Ray Midgley, Krister Mild, Christoph Moning, Dimitri Mouton, Killian Mullarney, Ray O'Reilley, Tommy Pedersen, Knud Pedersen, Colin Richardson, Gigi Sahlstrand, Dave Sargeant, Karen and Graham Searle, Michael Searle, Andrew Spalton, Karen Stanley-Price, Lars Svensson, Neil Tovey, Robert Tovey, Simon Tull, Magnus Ullman, Mattias Ullman, Reg Victor, Frank Walker, Ian Wallace and Effie Warr, plus hundreds of other dedicated birders who submitted their records to the Oman Bird Recorder over the years.

We would also like to thank those observers who provided comments on identification, regional status, and habitats in *Birds of the Middle East* (and *Birds of the United Arab Emirates*) which have been extensively used in our field guide. For these, special thanks go to Per Alström, Mike Blair, Oscar Campbell, Philippe Dubois, Dick Forsman, Steve James, Lars Jonsson, Mike Jennings, Nick Moran, Dick Newell, Tommy Pedersen, Roger Riddington, Abdulrahman Al-Sirhan, Andy Stoddart and Lars Svensson. Thanks are also due to the Natural History Museum (Tring) where we have examined skins over many years, and to the three artists, John Gale, Mike Langman and Brian Small for their ornithological advice and consummate professionalism in the production of the plates. In the quest for information and travels in the Middle East, over many years, BirdLife International, The Darwin Initiative, the Ornithological Society of the Middle East and the Wetland Trust have always given their support.

The preparation of the maps and status sections was greatly assisted by advice from Roger Brownsword, Gary Brown (Kuwait), Jamie Buchan (Qatar), Mike Jennings (Arabia), Howard King (Bahrain), Mike Pope (Kuwait) and David Stanton (Yemen).

Finally we would like to thank Jim Martin, Nigel Redman and Alice Ward at Bloomsbury Publishing for their unstinting support and advice; Julie Dando at Fluke Art for the preparation of the layouts; and Richard's co-authors of the first edition of the Middle East field guide, Per Schiermacker-Hansen and the late Steen Christensen and, in particular, Richard's co-author of the second edition (and the Emirates guide), the late Simon Aspinall.

All that now remains is to wish everyone safe travels and good birding.

Jens Eriksen and Richard Porter

INTRODUCTION

This field guide to the birds of Oman is an updated abridgement of the second edition of *Birds of the Middle East* (Porter & Aspinall 2010). Full coverage is given for all species, some 528 in total, known to have occurred in the wild in Oman up to June 2017. This total includes a number of naturalised species, with an additional selection of regularly observed free-flying escapes also being illustrated and described. Every year additional species find their way onto Oman's national checklist, the Oman Bird List, some being long expected, others coming as complete surprises. Indeed, many new species for the Middle East result from sightings made in Oman alone.

About 120 species of birds breed annually in Oman, the breeding avifauna being a unique blend of Western Palearctic, Oriental and Afrotropical components, the latter especially in the south of the country. Of the 528 species on the national checklist, more than 400 occur solely as migrant visitors, with more than 140 of these being especially rare, so-called 'vagrants'. This exceptional diversity is the result of the country lying on a migratory crossroads, with western and eastern flyways coming into contact with each other and having no clear divide. Unsurprisingly, Oman is a popular destination for visiting birdwatchers, joining a small but growing number of nationals and resident expatriates also pursuing this most absorbing hobby. Hopefully this field guide will both enable and encourage many more people to enjoy identifying those birds that they encounter, whether in city, park or garden, on the coast, or in the desert or mountains. Additional recommended reading dealing with Oman and its birds, and also with Middle Eastern birds in general, is provided under 'References and Further Reading' (pages 248–250).

A guide to the best birding sites is beyond the scope of this book and birdwatchers are recommended to consult the *Birdwatching Guide to Oman*: Second edition (Sargeant & Eriksen 2008) which covers all the best sites, and provides maps for every locality and much practical advice. More detailed information of

Crowned Sandgrouse, Mudday, southern Oman (Hanne & Jens Eriksen)

the status and distribution of birds in Oman can be found in the *Oman Bird List: Edition 7* (Eriksen & Victor 2013). Birdwatchers are specifically directed to the Oman Bird Recorder's website (www.birdsoman.com), which provides information on the latest bird sightings and various links to more information, as well as a comprehensive gallery of bird photographs and contact details for the Recorder. The Oman Bird Records Committee (OBRC) through its Recorder maintains the Oman Bird List and database, collating all observations submitted and assessing claims of vagrants. The definition of a vagrant varies between countries: in Oman the OBRC uses a minimum of 10 records as the 'cut-off' point above which a species ceases to be considered a vagrant, but as a rare visitor instead.

Around the world, birds are trapped and transported across borders for the cagebird trade, many subsequently either escaping or being released. In this respect the Middle East is no exception, and in recent decades several escaped species have established naturalised (self-sustaining) breeding populations in the wild in the region, notably parakeets, mynas and weavers. All such species deemed naturalised in Oman at the present day, of which there are a growing number, are included in this book. A selection of others from a long list of those not yet naturalised (some perhaps unlikely to become so), but which may nonetheless be observed free-flying, are also included here. Many are large and long-lived. Regrettably, the establishment of exotic species has sometimes had an adverse impact on the region's indigenous avifauna.

The *Birds of the Middle East* covered the following countries: Bahrain, Cyprus, Iran, Iraq, Israel, Jordan, Kuwait, Lebanon, Oman, Qatar, Saudi Arabia, Syria, Turkey, Palestinian territories (West Bank and Gaza), the United Arab Emirates (UAE) and Yemen, including the Socotra archipelago. We have retained substantially the same distribution maps from that publication in this guide, since it is helpful to see the bigger, regional picture, and one can see at a glance which species occurs where and when (if resident or migrant, and if the latter whether it breeds or not, and so on). To the best of our knowledge, only species accepted onto a checklist by the relevant national body in any country are included in the maps or status text.

Verreaux's Eagle, Dhofar Mountains (Hanne & Jens Eriksen)

TAXONOMY, NOMENCLATURE AND SEQUENCE

This is a very confusing and contentious area of ornithology, and newcomers to birding will have every reason to be mystified by the various treatments given by different books. There are those taxonomists who prefer to 'lump' species and those who prefer to 'split' them (thus subspecies become species in their own right).

The species taxonomy and nomenclature (English and scientific names) followed here is that of the list published by the International Ornithological Congress (IOC) and adopted by the Ornithological Society of the Middle East for its 'OSME Region List' (ORL). The IOC list is updated every quarter and can be viewed online at www.worldbirdnames.org while the OSME list can be seen at www.osme.org/orl. For presentational reasons, the sequence of families and species does not follow the IOC list precisely. See the Checklist on pp.251–261 for the latest official order of species.

Only rarely have English names deviated from those proposed by the IOC, in each case involving logical divergence from that proposed by the IOC, rather than (and despite temptation) any particular personal preference. We have also given alternative names where relevant (under 'Alt' at the end of the species text), but only those that are still often used or are of help in preventing confusion. The treatment given to alternative names is not comprehensive but if a problem arises, the scientific name should prevent ambiguity, although many of these have also been subject to change in recent years.

In vogue with the times, though not without substantial scientific backing, the taxonomic trend has been towards splitting species and the recognition of new species. With such a constant state of flux, it is impossible to stay ahead of developments. This field guide is not a taxonomic authority and has simply embraced the latest thinking on what may or may not constitute a species. However, we describe all recognisable taxa known to occur in the region.

Some explanation of the use of binomial and trinomial scientific names is required:

- Where there is universal agreement on full species status a binomial is used in the scientific name, e.g. *Merops orientalis* (Green Bee-eater).
- For taxa which are universally recognised as different forms (subspecies) of the same species then the trinomial is used without brackets, e.g. *Buteo buteo vulpinus* (Steppe Buzzard).
- We have generally avoided using parentheses in any trinomial names following the recommendations of the IOC list. Several subspecies, for example, *Acrocephalus stentoreus brunnescens* (Indian Reed Warbler) may possibly be upgraded to full species status, e.g. *Acrocephalus brunnescens*, in the future.

ILLUSTRATIONS AND IDENTIFICATION TEXT

Our aim has been to make the illustrations and species accounts of help to beginner and expert alike. We have concentrated on those features which are important for identification and, accordingly, these are highlighted in the text. For each species, length from bill tip to tail tip (L) is given in centimetres; for larger birds wingspan (W) is also given. Where the identification of a species does not present a problem, the texts are often brief, whereas more difficult species have necessarily warranted more detailed descriptions, sometimes including biometrics.

There is no need to stress the increasing importance of digital photography in helping to make correct identifications, especially of tricky species. This enables examination and discussion with experts at a later date.

Abbreviations used on the plates: ad – adult, juv – juvenile, imm – immature, ♂ – male, ♀ – female.

BIRD TOPOGRAPHY

The illustrations below show the various features of a bird – its topography – used in the identification texts. Knowledge of these, especially the feather tracts, is vital for describing a bird and its plumage. Making field sketches, even if embarrassingly poor, is a good way to learn.

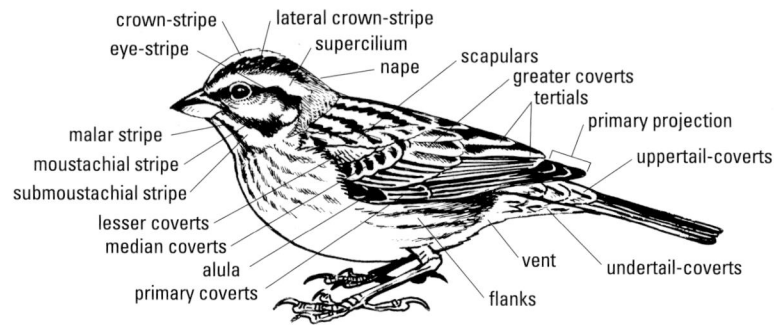

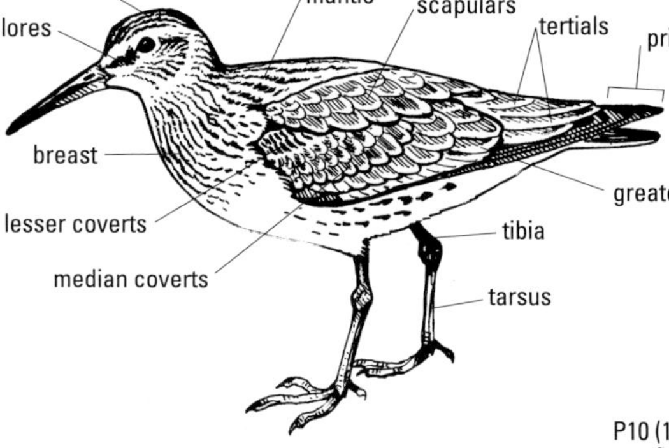

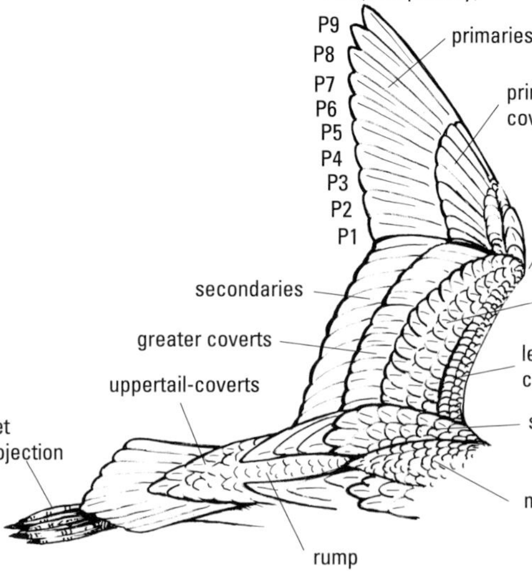

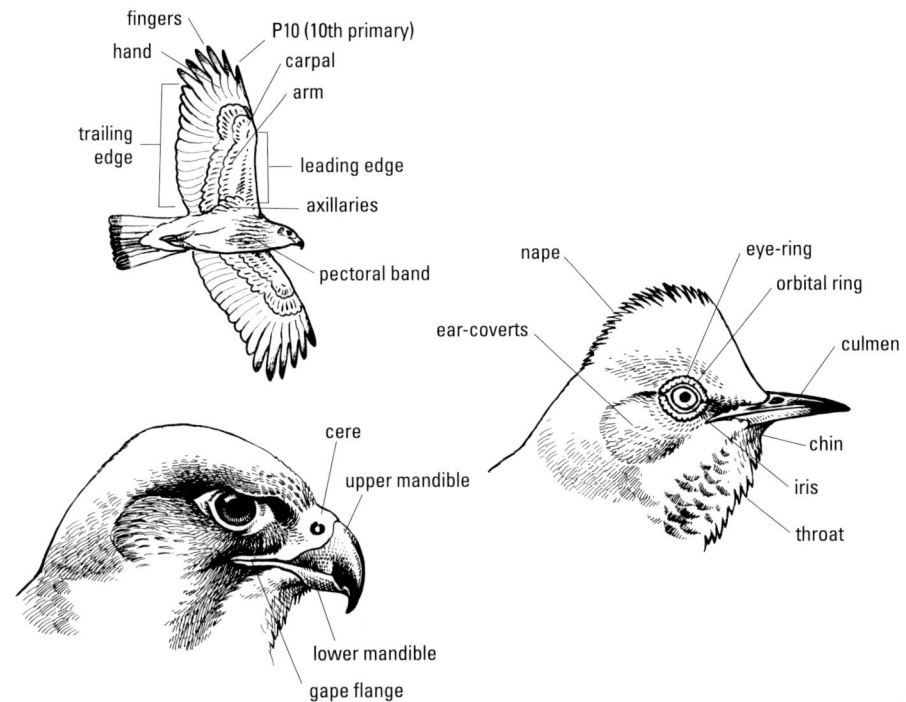

VOICE

Calls and song are only given where a species occurs regularly in the region or where it might be helpful for identification. Song is of course given for all species breeding in Oman, but many migrants also give at least occasional bursts of song or sub-song on passage or in winter, even though moving on to breed elsewhere. Where this is known to be the case in Oman then description of those songs is also given.

A large number of sound guides are now widely available, and given the limitations of phonetic transcription we strongly recommend that one be purchased. The internet can also be checked for sound recordings. There is no substitute for learning vocalisations or, if unable to commit them to memory, in comparing recordings. However, use of the technology to record calls and songs has somewhat lagged behind digital photography as a means for confirming the identity of a bird. Few of us use sonograms, but the way forward certainly lies in this field. In our region, for example, the calls of the different subspecies of Common Chiffchaff *Phylloscopus collybita* may only be separable by analysis in this way. Indeed, many recently described new species of birds (and even some mammals) have been first noticed as such on the basis of their different vocalisations.

There is also, as amply demonstrated by our own species, *Homo sapiens*, the matter of differing regional dialects. Songs and calls of bird species commonly found in Western Europe may often sound quite different in the Middle East. Even within the region, strong intraspecific variations can frequently be noticed.

HABITAT

The habitats described in the text are those occupied in the Middle East and this may help in species identification or at least narrow down the list of possible contenders, or direct you where to look for a particular species. Habitats occupied by species visiting Oman in winter and/or on passage are detailed when differing from those utilised in their Middle Eastern breeding grounds.

MAPS AND STATUS

As a quick guide, the status of each species in Oman appears to the right of the species header, using the codes given below:

RB/rb – Resident breeder
MB/mb – Migrant breeder
PM/pm – Passage migrant
WV/wv – Winter visitor (Dec–Feb at least)
SV/sv – Summer visitor (non-breeding)
V – Vagrant (fewer than 10 records)
E/I – Escape or Introduction

Upper and lower case letters are used to indicate relative abundance of that species, upper case reflecting larger numbers or greater abundance either at that season and/or in comparison with other closely related species. When PM and WV (or pm and wv) share the same case the season when that species is the more common will appear first, unless they are actually equally common at the two seasons.

The regional distribution maps distinguish between resident and migrant breeders and also show the passage and/or winter range for all regularly occurring species. Vagrants are not mapped. Note that the maps show current known ranges (not historical ones). Further details on status in the Middle East, to complement the maps, are given as necessary under '**Note**' at end of the species accounts.

Broad geographical divisions are sometimes used:

Near East – the eastern Mediterranean countries (excluding Turkey and Cyprus)
SW Arabia – Yemen and adjacent areas of Saudi Arabia
SE Arabia – Oman, UAE and adjacent areas of Saudi Arabia (east of the Qatar peninsula)

The breeding, passage and winter distributions of birds are now relatively well known for much of the Middle East, although gaps still exist. Regrettably, it has not been possible at this time, even when appropriate, to map passage and winter ranges separately. In any case, in the south of the region, where the seasons are not clear-cut, autumn migration may continue, for example, until December, while spring migrants may reappear as early as January. In general, however, a species shown by hatching across the entirety of Arabia will generally be a passage migrant that is absent in winter or present then only in relatively low numbers.

Note that many non-breeding waterbirds, often immature birds, may be found year-round on the coast or at inland wetlands, including in particular many different shorebirds, amongst them even arctic-breeding species. Do not be surprised to see, for example, Ruddy Turnstone *Arenaria interpres* or Sanderling *Calidris alba* on the foreshore in Oman in mid-summer.

Some species mapped as breeding visitors may actually be resident in some areas, usually more southerly localities, and/or perhaps involving only a small proportion of the population. (Interestingly, some classic Palearctic–African migrants are now found wintering in newly vegetated areas of Arabia.) Moreover, as the Tropic of Cancer traverses Arabia it is perhaps hardly surprising that the breeding season is not always the spring and summer; winter breeding is the norm for some species, including migrant species which elsewhere otherwise breed conventionally in summer. The breeding season of other species in Dhofar (southern Oman), including some migrants from Africa, is regulated by the late summer monsoon and occurs in the 'autumn' months.

Extensive use has been made of the excellent maps prepared for the *Atlas of the Breeding Birds of Arabia*, to which many individuals have contributed, and of the country avifaunas listed in 'References and Further Reading' (pages 248–250). The mapping scale and accuracy varies from one country avifauna to another, and at times we have had to extrapolate or interpret to the best of our combined knowledge those maps produced with, say, a 'broader brush' approach.

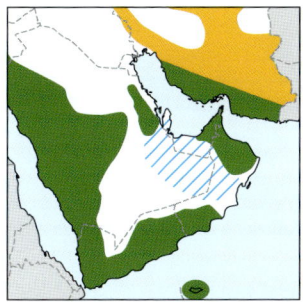

Green – Resident (present all year)

Orange – Migrant breeder (seen in the breeding season and on passage)

Blue hatch – Passage and/or winter visitor

CODE OF CONDUCT AND SUBMITTING RECORDS

The welfare of birds must come first, particularly of nesting birds. Many ground-nesting species may not incubate by day, merely shade their eggs, and prolonged disturbance may result in the embryos being baked alive, or in nest desertion. Try to minimise time spent near nests or chicks; even if you haven't actually located them it is usually evident from the behaviour of the adult(s) that are nearby. Photographers please take note! Roosting and feeding birds, notably waders, are easily disturbed so please be thoughtful when, for example, trying to get that frame-filling shot of a Crab-plover *Dromas ardeola*.

We all now recognise the importance of digital photography in helping to make correct identifications of tricky species and also, all-importantly, in convincing national records committees of any rarer sightings that you may make. On this point, please do remember to submit your observations to the national recorder – all may be helpful in promoting ecotourism, lobbying for conservation and in bringing added protection to important sites and species. The relevant body in Oman for submission of your sightings is the Recorder of the OBRC (see www.birdsoman.com for contact details).

Finally, please observe cultural and religious sensitivities in the region. Shorts and sleeveless T-shirts are not appreciated, especially when worn by women. Hospitality shown to visitors is second to none, but be aware of and observe local customs and etiquette. Always accept a cup of tea when offered, unless you are going to miss your flight home, and even if you don't like sugar.

Jouanin's Petrel, off Mirbat (Hanne & Jens Eriksen)

PLATE 1: WHISTLING DUCKS AND SWANS

Fulvous Whistling Duck *Dendrocygna bicolor* V

L: 50. W: 80. Shares upright stance and long neck and legs with similar but smaller Lesser Whistling Duck; told from latter by *larger, warm orange head with longer bill giving elongated profile, and dark line down hindneck; also dark streaking on sides of neck and prominent white streaks on flanks*. In rather weak flight both whistling ducks show rather rounded wings with *dark upper- and underwings* and deep wingbeats. **Habitat** Freshwater wetlands. **Note** Vagrant Oman, Yemen, from Africa or Orient.

Lesser Whistling Duck *Dendrocygna javanica* V

L: 42. W: 70. Similar to but smaller than Fulvous Whistling Duck, which see, being told by *greyish-buff head and neck, lacking dark line down hindneck, and absence of black streaking on neck-sides*. In flight note bright chestnut forewing and uppertail-coverts. **Habitat** Freshwater wetlands. **Note** Vagrant Oman, from Orient.

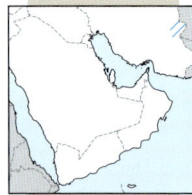

Mute Swan *Cygnus olor* V

L: 153. W: 223. *Orange bill with black knob* in adults (larger in male). *Typically with graceful S-curve to long neck, bill often pointing down*. Juvenile dingy-brown with *black-based grey bill* gradually becoming pink and then orange. Plumage grows increasingly white during first winter and spring, sometimes still partly brownish above until second winter. *In flight, wingbeats produce loud, rhythmic, singing sound vaou-vaou-vaou*. **Voice** Mostly silent. **Habitat** Lakes, marshes, deltas; in winter also sheltered sea coasts; builds large nest near water. **Note** Winter hatched; vagrant Gulf States, Oman.

Whooper Swan *Cygnus cygnus* V

L: 153. W: 231. Similar to Bewick's Swan, but larger (size of Mute Swan) with proportionately longer neck and bill; *black bill with prominent yellow base extending in wedge to below nostril or beyond*. Flight silent but with powerful wingbeats like Mute Swan. *Juvenile/immature similar to Bewick's, but note longer neck and wedge-shaped head profile*. Young birds often with parents during first winter. **Voice** Similar to Bewick's but deeper and stronger and with musical, trumpet-like quality *ahng-ha* or *ko-ko-ko*; often trisyllabic. **Habitat** Tidal waters, lakes, rivers and floodplains, fields. **Note** Vagrant Oman, Qatar, UAE.

Bewick's Swan *Cygnus columbianus bewickii* V

L: 121. W: 195. *Smallest swan, with shorter neck and bill* than Mute and Whooper Swans. Similar to larger Whooper (both having straight neck), but *yellow on bill reduced to base, ending well behind nostril*. Immature greyer than same age Mute Swan, and *pink bill lacks black base*, becoming white and later yellow during first winter. From immature Whooper by *shorter neck and bill and more rounded head*. In flight wingbeats slightly faster, more goose-like and with no humming sound. **Voice** Similar to Whooper but higher pitched, monosyllabic or disyllabic (Whooper often has three syllables), sometimes recalling distant barking dogs. **Habitat** Inland and coastal wetlands. **Note** Winter hatched; vagrant Oman, Saudi Arabia, UAE.

Knob-billed Duck *Sarkidiornis melanotos* V

L: 76. W: 150. Large duck, with strong flight recalling goose. *Male has large fleshy knob on top of bill and forehead*, absent in female and juvenile. *White head and neck speckled blackish*, underparts largely white, in male with narrow black half-collar down sides of breast (absent in rather smaller female which has also more mottled underparts). Upper back blackish, glossed green and purple, *lower back conspicuously grey*. In flight, upperwing, rump and tail appear all-dark except for bronze speculum. Immature duller, less glossy than female. Perches freely on trees; feeds largely by grazing. **Habitat** Wetlands. **Note** Vagrant Oman, from Africa or Orient. [Alt: Comb Duck]

PLATE 2: GEESE

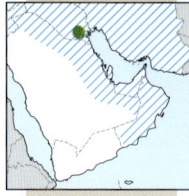

Greylag Goose *Anser anser* wv
L: 83. W: 164. Only the paler eastern subspecies *rubrirostris* (**Eastern Greylag Goose**) has been recorded in the region. A large goose with thick neck, *heavy, pink bill, pale greyish head and neck, and pink legs*. In flight, shows distinctive *pale grey forewing*. Juvenile similar to adult but lacks sharply defined transverse lines of upperparts and dark belly marks. From all other geese in the region by combination of size, bill and leg colour and strikingly pale forewing. **Voice** Similar to domestic goose; loud, characteristic *ang-ang-ang* in flight. **Habitat** Grasslands, arable fields, marshes, estuaries; breeds in marshes, reedbeds, boggy thickets, islets. **Note** Passage and winter hatched.

Greater White-fronted Goose *Anser albifrons* wv
L: 72. W: 148. Smaller than Greylag, and normally larger than Lesser White-fronted Goose. Warmer grey-brown than Greylag, with slightly darker head and hindneck; has *orange legs* and lacks silvery forewing. Wing pattern similar to that of Lesser White-fronted. Bill not heavy, but quite long, *pink with white nail;* no eye-ring visible in field. Adult has large white area surrounding base of bill (rarely also forecrown) and *black bars on underparts*. Juvenile browner, lacks white forehead and black bars on underparts; *forehead and area round bill very dark, bill has dark nail*. **Voice** Musical and much higher pitched than other grey geese except Lesser White-fronted; repeated disyllabic, sometimes trisyllabic with metallic, laughing quality *kow-lyow* or *lyo-lyck*. **Habitat** Grasslands, marshes, estuaries. **Note** Passage and winter hatched; vagrant Kuwait, Saudi Arabia, Yemen.

Lesser White-fronted Goose *Anser erythropus* V
L: 59. W: 127. Resembles Greater White-fronted Goose, but smaller (though a few overlap in size), with proportionately *shorter neck and bill, steep forehead, yellow eye-ring* and slightly longer wings, often extending beyond tail. Adult has *extensive white on forehead and forecrown to above the eye*; upperparts dusky-brown with only dull transverse lines; underparts with only a few black blotches. Juvenile similar to juvenile Greater White-fronted Goose, but told by size, head and neck proportions and yellow eye-ring. Compared to Greater White-fronted has faster wingbeats, more compact silhouette and noticeably faster walk. **Voice** Higher pitched, squeakier and faster than Greater White-fronted; a di- or trisyllabic yelp *kow-yow, kyu-yu-yu* or piping *yi-yi-yi*. **Habitat** Grasslands, marshes. **Note** Winter hatched; vagrant Kuwait, Oman, UAE.

Red-breasted Goose *Branta ruficollis* V
L: 55. W: 125. Small, compact, dark goose with *short, black bill*. Unmistakable pattern of *black body, chestnut cheeks and breast* (looks dark at distance) *bordered by white*; broad, shaggy, white line along flanks and white patch in front of eye. Juvenile has smaller chestnut cheek-patch. Often mixes with the so-called 'grey geese'. Flight fast and agile. **Habitat** Normally grassy steppes in winter. **Note** Winter hatched, but rare; vagrant Cyprus, Israel, Oman, Syria.

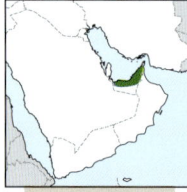

Egyptian Goose *Alopochen aegyptiaca* V
L: 68. W: 144. Slightly larger than Ruddy Shelduck, with broader wings, but similar wing-pattern in flight. Otherwise *brownish-buff* with brownish or greyish upperparts, pale foreneck and face, and *prominent dark eye-patch*; dark collar and belly-patch. Juvenile/immature browner, lacking eye-patch, collar and belly-patch, and with yellowish-grey legs. Told from Ruddy Shelduck by longer legs, dark eye-patch and body colour. **Voice** Deep nasal braying. **Habitat** Freshwater marshes, lakes and rivers with nearby grassland. **Note** Vagrant Oman, introduced UAE.

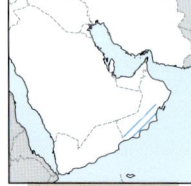

Cotton Pygmy Goose *Nettapus coromandelianus* wv
L: 33. W: 55. Smallest duck to occur in region. Short neck and short, stubby, goose-like bill; round body with, on water, high rear end. Male told by white head, neck and underparts with black cap, eye and band across white breast. In fast flight looks white with green-tinged black wings with conspicuous white trailing edge across full length of wing, very broad on primaries. Female a drab version of male, browner with dark eye-line and greyish flanks; in flight brown wings and narrow white trailing edge to secondaries. **Habitat** Well-vegetated wetlands, lakes. **Note** Uncommon but regular in S Oman; vagrant Bahrain, Qatar, Saudi Arabia, UAE, Yemen (Socotra), Iran and Iraq. [Alt: Cotton Teal]

PLATE 3: SHELDUCKS AND DABBLING DUCKS I

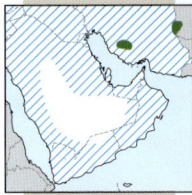

Common Shelduck *Tadorna tadorna* wv
L: 63. W: 122. Large with white body, *black head with green gloss, broad chestnut band around breast and prominent red bill with fleshy knob in breeding male*. In flight, black flight feathers contrast with white upper- and underwing-coverts and black shoulders. Juvenile lacks chestnut band, has grey-brown head, hindneck and upperparts, and pale greyish bill. **Voice** Male in breeding season has high whistling *siss-siss-siss*, answered by the female's whinnying *gehehehehheh*; sometimes nasal *ah-hang*. **Habitat** Sandy and muddy coasts, fresh and saltwater lakes and marshes; nests in burrow, hole in tree or under thick cover. **Note** Passage and winter hatched, but uncommon in much of Arabia.

Ruddy Shelduck *Tadorna ferruginea* wv
L: 64. W: 133. Size of Common Shelduck with *orange-chestnut body* and paler cinnamon-buff head; *black flight feathers* with green speculum, *striking white forewing and underwing-coverts* (similar to Egyptian Goose). Female and non-breeding male lack black neck collar; female also has paler head. Juvenile like adult female, but back browner. From Egyptian Goose by shape, narrower, more pointed wings, head pattern and darker chestnut underparts. **Voice** Nasal, trumpeting and rather penetrating *ang, ang*. **Habitat** Sandy lake shores, river banks, fields, arid steppe; nests in hole or burrow. **Note** Passage and winter hatched, but rare or irregular in S Arabia.

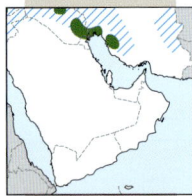

Marbled Duck *Marmaronetta angustirostris* V
L: 41. In flight resembles small female Northern Pintail, having fairly long neck, wings and tail. On the water identified by *pale plumage, dappled dark and cream* (marbled) with *dark oval patch around eye*. Head large and rounded with steep forehead and, in adult, bulky crest on lower nape. In flight, wings are fairly pale with slightly paler secondaries but *no speculum*. Often secretive, hiding in vegetation. **Voice** Low nasal wheezing *jeak* or double whistling note. **Habitat** Well-vegetated lakes and reservoirs; also salt lakes in winter. **Note** Winter hatched but mostly rare; vagrant Bahrain, Cyprus, Kuwait, Oman, Qatar, Saudi Arabia, UAE.

Indian Spot-billed Duck *Anas poecilorhyncha* E/I
L: 60. W: 100. Large, rather pale duck with *yellow tip to dark bill* and dark line through eye on otherwise pale grey face and neck; male has reddish spot on lores. Upperparts and flanks boldly scalloped; breast spotted dark. *Black undertail-coverts noticeable, particularly when up-ending. In flight, shows conspicuous white tertials*. Size and shape as Mallard. **Habitat** Wetlands. **Note** Introduced Salalah, Oman in 1990s, surviving for several years; several unconfirmed records in recent years.

Gadwall *Anas strepera* wv
L: 51. W: 89. Medium-sized, slightly built dabbling duck. Flies with rapid wingbeats and pointed wings like Eurasian Wigeon. *Male mainly dark grey with black around tail; in flight, white speculum bordered black in front, and white belly*. Female and juvenile resemble larger and heavier female Mallard, but tail grey-brown (not white), and thin dark bill is orange along sides; *in flight, shows white speculum patch close to body* (often smaller in male) and *white belly* bordered by dark flanks. **Voice** Male has a rasping, low *rrep* call. Female's call resembles Mallard's but is higher pitched. **Habitat** Any wetland. **Note** Passage and winter hatched.

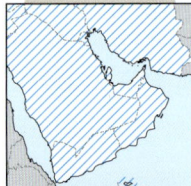

Eurasian Wigeon *Anas penelope* wv
L: 48. W: 80. *Rather rufous duck* with short neck, rounded, *steep forehead and small blue-grey bill. In all plumages has white belly-patch*, obvious in flight and when grazing. Adult *male has creamy-yellow forehead*. Adult female rufous to greyish-brown; lacks white forewing. Juvenile resembles adult female (immature male does not usually assume white forewing until second winter). In flight, in all plumages shows narrow wings, pointed tail, and contrasting white belly-patch; *white forewing contrasting with blackish-green speculum is distinctive in adult male*. Female and juvenile have grey-brown forewing, darker flight feathers and black speculum. **Voice** Male has characteristic clear whistle *wheeooo*, both in flight and on water; female growls in flight. **Habitat** Coastal mudflats, marshes, lakes. **Note** Passage and winter hatched.

PLATE 4: DABBLING DUCKS II AND MERGANSER

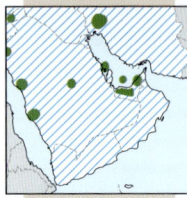

Mallard *Anas platyrhynchos* WV
L: 56. W: 95. Large dabbling duck. Male has *grey wings, dark blue speculum distinctly bordered by white, and white underwing-coverts*. Female mottled brown like other female ducks and told by size, bill shape and colour, and wing pattern. Male in eclipse resembles female, but bill greenish-yellow. **Voice** Male has a soft nasal *raehb* and, during courtship, a weak, high-pitched whistle *piu*. Female gives familiar deep quacking. **Habitat** Any wetland, including estuaries in winter. **Note** Arabian population feral; passage and winter hatched.

Northern Shoveler *Anas clypeata* WV
L: 51. W: 78. *Huge, spatulate bill*. Swims with front end low and bill often dabbling in water. In flight, wings appear set far back. Male unmistakable; in flight, looks black, white and chestnut with *distinctive blue forewing*. Female and juvenile on water resemble female Mallard and Gadwall, but bill distinctive; in flight, shows bluish forewing, somewhat resembling smaller male Garganey. Male in eclipse largely resembles female, but brighter blue forewing. **Voice** Male calls a hollow, double-note *g-dunk, g-dunk*, often in flight, while female simultaneously quacks *pe-ett*. **Habitat** Marshes, lakes, ponds. **Note** Passage and winter hatched.

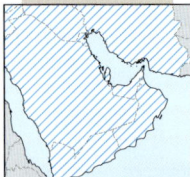

Northern Pintail *Anas acuta* WV
L: 56 (excluding long tail feathers of male). W: 88. Slim and elegant dabbling duck. Male has *white neck and underparts with dark head and tail and dark slender bill*. Female recalls other female dabbling ducks, but note *dark bill, greyer plumage and much slimmer appearance*. In flight both sexes show distinctly *longer neck and tail*, and slender, more pointed wings than other ducks; also note *green speculum with conspicuous white border to rear*. Female has distinctive *white border at rear of secondaries*; upperwing otherwise brownish. **Voice** Male has a low-pitched, weak whistle; female a hoarse quack. **Habitat** Sheltered coasts and estuaries in winter; also shallow inland waters. **Note** Passage and winter hatched.

Garganey *Anas querquedula* PM, wv
L: 39. W: 63. Small dabbling duck. *Male shows long white stripe on head and, in flight, striking blue-grey forewing*. Female similar to slighter female Eurasian Teal, but *longer, heavier bill and more contrasting dark and light head-stripes, widening in front into pale patch at bill base*, and whiter throat. In flight, female's forewing is slightly paler and white border along secondaries is distinctive (Eurasian Teal has white wing-bar in middle of wing in front of speculum). Male in eclipse is like female, but wing pattern as adult male. Juvenile similar to female. **Voice** Male calls a dry, drawn-out rattled *knerreck*; female a short, sharp quack. **Habitat** Freshwater wetlands. **Note** Passage hatched; some winter in Arabia.

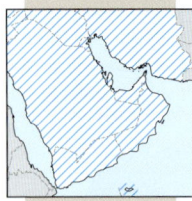

Eurasian Teal *Anas crecca* WV
L: 36. W: 61. Commonest small duck; readily takes to the wing, both sexes showing white belly and *greyish underwing with light band through middle*. Adult male distinctive; female rather similar to Garganey (which see for differences). Often in fast-flying, tightly-knit twisting and turning flocks. **Voice** Male has far-carrying ringing whistle *kreek-kreek*; female a high-pitched, nasal quacking. **Habitat** Wetlands, from saltmarshes to lakes and ditches. **Note** Passage and winter hatched.

Red-breasted Merganser *Mergus serrator* V
L: 56. W: 78. Conspicuous ragged wispy crest in both sexes. In fast direct flight, appears elongated with long head and neck, and shows much white on inner wing. Male's greenish-black head, thin red bill, white neck collar and rusty black-spotted breast easily separate it. Female and juvenile closely resemble female Goosander *M. merganser*, which is unrecorded in Oman, but smaller and slimmer-billed, with more brownish (less greyish) upperparts, less contrasting head pattern, and without sharp demarcation between brown head and greyish-buff neck and breast; flanks darkish grey. **Habitat** Chiefly maritime. **Note** Winter hatched, often rare; vagrant in The Gulf to Oman.

PLATE 5: DIVING DUCKS, GUINEAFOWL AND BUTTONQUAIL

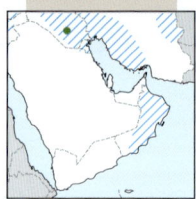

Red-crested Pochard *Netta rufina*   wv
L: 56. Large diving duck which sits high on the water. Male unmistakable *with large dark orange head and red bill*; in eclipse resembles female but bill is red. Female has pale grey cheeks contrasting with dark crown; dark bill has pink band near tip. In flight shows broad white wing-bar. **Voice** Male has a double *weep-weep*; female a grating *keerr*. **Habitat** Freshwater lakes, estuaries. **Note** Winter hatched, uncommon Oman; vagrant Bahrain, Qatar, Saudi Arabia, UAE.

Common Pochard *Aythya ferina*   wv
L: 45. W: 80. *Sloping forehead grading into long bill* is characteristic. Male has *chestnut head and neck, contrasting with black breast*; black bill has pale blue-grey band in centre. In eclipse resembles female but greyer above. Female has dull brownish head and breast with paler chin and eye-stripe; dark bill becomes paler towards broad black tip. In flight looks longer and plumper than Tufted Duck; greyish-brown wings (paler in male) have *indistinct pale grey wing-bar*. **Voice** Courting male has low whistle; also hoarse wheezing note; female has rough harsh *krra-krra*. **Habitat** Well-vegetated wetlands. **Note** Passage and winter hatched.

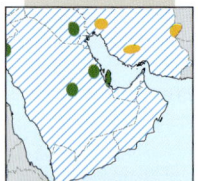

Ferruginous Duck *Aythya nyroca*   wv
L: 40. W: 66. Slightly smaller than Tufted Duck, head shape close to Common Pochard. *Compared to Tufted Duck has higher crown, flatter sloping forehead and longer, dark grey bill with black nail only*. Male rich chestnut-brown with white eye and sharply defined white undertail-coverts. Female and immature have dark eyes (but white in one-year-old male); they lack whitish at base of bill of some Tufted Ducks; *female has warmer brown head than Tufted* (which has yellow eye); pure white and sharply defined white undertail-coverts (sometimes seen in Tufted). *First-year birds told by head shape, length and pattern of bill, and dark eye*. In flight, *broader, more conspicuous white wing-bar than Tufted Duck*. **Voice** In flight, high-pitched *crr-err*. **Habitat** Shallow, well-vegetated wetlands. **Note** Passage and winter hatched, mostly scarce. Occasionally summers in Oman.

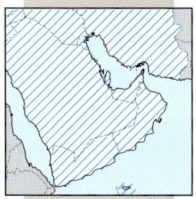

Tufted Duck *Aythya fuligula*   wv
L: 42. W: 70. Small diving duck with roundish head but fairly steep forehead; *crest on nape*, long and drooping in male, minute in female. *Bill blue-grey* (darker in female) *with broad black tip*. Male has black upperparts and purple head sheen. Female and male in eclipse have brownish sides to body; female sometimes with whitish band around base of bill. In summer, female has *darkish back, short crest and broad black tip to bill*. Some females show white undertail-coverts. In rapid flight has conspicuous white wing-bar, almost the full length of wing. **Habitat** Lakes, ponds. **Note** Passage and winter hatched.

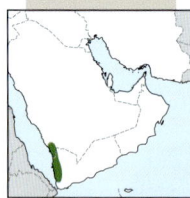

Helmeted Guineafowl *Numida meleagris*   E/I
L: 63. W: 97. Size of domestic chicken. Note *horny, brown protuberance on crown*. Juvenile brownish with buff and white spotting and less pronounced head pattern. Gregarious, runs fast with head held high or flies rapidly with occasional glides. **Voice** Far-carrying, raucous *kek, kek, kek, kek, kaaaaaa, ka, ka, ka, ka, kaaaaa, ka ka*, uttered by flocks or at roost. **Habitat** Savanna, wadis, scrub, thick cover. **Note** Records of escapes from captivity from N and S Oman.

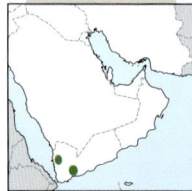

Common Buttonquail *Turnix sylvaticus*   V
L: 15. W: 27. Very hard to flush. Resembles Common Quail in buffish plumage, shape and gait, but smaller with *uniform orange breast with bold black spots on sides; lacks solid dark head marks of Common Quail*; female brighter than male. *In flight, lark-sized with short tail, short, rounded wings* (quite unlike the quails, which have narrow wings), *dark flight feathers contrasting with buffish wing-coverts and body*; rapid, whirring wingbeats for short distance before settling. **Voice** Female has curious deep, hollow, resonant *crooo…* or *hooo…* usually lasting one second and repeated every 1–2 seconds for 30 seconds or more; often at night and resembles distant lowing of cattle. **Habitat** Sandy plains with palmetto scrub, brush-covered wastes, neglected cultivation. **Note** May breed Iran, but rare; vagrant Oman. [Alt: Andalusian Hemipode, Kurrichane Buttonquail]

PLATE 6: PARTRIDGES, FRANCOLIN AND QUAILS

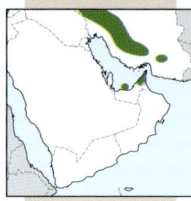

Chukar Partridge *Alectoris chukar*   RB
L: 33. A distinctive plump gamebird; *grey-brown with striped flanks. Bill and legs red*. Wary, runs fast; usually in small flocks. **Voice** Characteristic accelerating *ka-ka-kaka-kaka-kaka...* followed by a chuckling *chukara-chukara-chukara-chukar*, frequent at dawn and dusk especially. **Habitat** Rocky slopes and hillsides, semi-deserts, agricultural land; from sea level to 3,000m, occasionally higher. **Note** Native to Musandam but not found elsewhere in Oman.

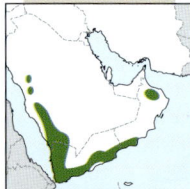

Arabian Partridge *Alectoris melanocephala*   RB
L: 36. Larger than Chukar Partridge but ranges do not overlap. Black crown and long, broad supercilium also separate it from Chukar. Juvenile, which lacks flank-bars and gorget, is grey-brown above, grey-buff below, and lightly pale-spotted all over. Most often in small flocks; usually escapes danger on foot. **Voice** Typically *kok, kok, kok, kok, kok, chok-chok-chok chook* beginning as separate notes, then accelerating and descending; also a rapid *chuk-chuk chuk chuk chuk*; a loud throaty *crook* and conversational, soft *croo, croo, croo, croo*. Often heard during the day. **Habitat** Rocky slopes with bushes, from sea level to 2800m. **Note** Common in S Oman, uncommon N Oman.

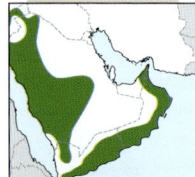

Sand Partridge *Ammoperdix heyi*   RB
L: 24. W: 40. Larger than Common Quail, but much smaller than Chukar Partridge. Runs nimbly over rocks; wings whirr noisily in flight. *Male has white patch on forehead and behind eye*, but lacks black head markings of extralimital See-see. Male of subspecies *intermedius*, occurring in Arabia, is dark pinkish-cinnamon above (fading sandier when worn); flanks heavily banded black and chestnut. Both sexes show chestnut outer tail feathers in flight. Female and immature are uniform sandy-grey. Often forms crèches, with large flocks arriving on foot to drink from small pools. **Voice** Commonest call is a repeated, metronomic, *qwei, qwei, qwei*; alarm call *wit-wit* in flight. **Habitat** Desolate, arid rocky and stony slopes, wadis and cliffs. **Note** Fairly common in Oman, but avoids Empty Quarter.

Grey Francolin *Francolinus pondicerianus*   RB
L: 30. Rather drab, stub-tailed, greyish-brown gamebird separated from slightly larger female Black Francolin by *much finer mottling and barring on upper- and underparts, with chestnut forehead and cheeks*. Lacks chestnut patch on hindneck and has *pale throat-patch bordered below by black 'U'*. In flight, shows chestnut tail. Not secretive, often feeding in small parties in the open. Usually runs from danger but, if pressed, rises with explosive whirr of wings. **Voice** Loud far-carrying series of 9–15 notes: *kik-kjyw-ku, kik-kjyw-ku, kik-kjyw-ku*, commonly heard. **Habitat** Scrub, edges of cultivation and semi-desert. **Note** Range expanding naturally following deliberate introduction.

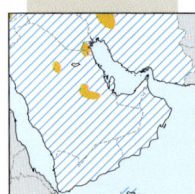

Common Quail *Coturnix coturnix*   PM, wv, mb?
L: 17. W: 33. Small gamebird, *more often heard than seen*, and hard to flush. In flight, *size of Common Starling*; rather pointed, narrow wings bowed, fast shallow wingbeats, *plain wings and striped back*. Creeps about on ground inconspicuously. Female has paler head pattern and lacks neck-band of male. **Voice** Characteristic sound in open country over much of the region, especially in farmland, by day or night; an explosive, rhythmically repeated *trisyllabic whistle pit, pil-it* (rendered as 'wet-my-lips'). **Habitat** Grasslands, cereal crops, meadows (sometimes in mountains). **Note** Passage hatched, rare in winter.

Harlequin Quail *Coturnix delegorguei*   V
L: 16. Slightly smaller than Common Quail, with plainer upperparts and *much darker underparts (male)*. Rarely seen unless flushed, when less narrow-winged than Common Quail. Male told by *broad black centre of breast, chestnut flanks with black streaking*, and black-and-white head pattern. Female has *unmarked buff throat and warmer rufous lower breast and belly with blackish spots* (though present in juvenile). Small Buttonquail looks much smaller in flight with short, rounded wings and contrasting upperwing pattern. **Voice** Resembles Common Quail but more metallic: an explosive *whit-whit-whit* or rising and falling *wit, wit-wit, wit-wit* repeated at intervals of 1–2 seconds. **Habitat** Grassland and cultivated fields. **Note** Apparently an irregular visitor but may breed SW Saudi Arabia and Yemen (where recorded in numbers in late 19th century); vagrant Oman, Socotra.

PLATE 7: STORM PETRELS

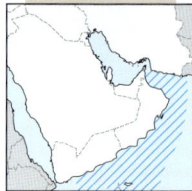

Wilson's Storm Petrel *Oceanites oceanicus* SV
L: 18. W: 40. Most numerous storm petrel in the region. Identified in flight by *bold white rump extending onto sides of uppertail-coverts, and legs protruding beyond square-ended tail*. Fairly conspicuous panel on the upperwing-coverts, with *underwing wholly dark. Webs of feet yellowish*, sometimes seen on pattering bird. Flight fluttering with series of flaps interspersed with short glides. Will follow ships, unlike most storm petrels. **Habitat** Maritime. **Note** Regular summer visitor in hatched area, from Southern Ocean; rare in The Gulf, vagrant Red Sea to Gulf of Aqaba.

Swinhoe's Storm Petrel *Oceanodroma monorhis* SV
L: 20. W: 45. *All-dark storm petrel, including rump*. Deeply forked tail. Larger than Wilson's Storm Petrel. Pale bar across upperwing-coverts not always obvious. Very similar to Matsudaira's Storm Petrel and separated from that species by *longer but narrower white shafts to base of outer primaries on upperwing*. Flight bouncing and Black Tern-like. **Habitat** Maritime. **Note** Rare summer visitor to Oman and UAE from Japan.

Matsudaira's Storm Petrel *Oceanodroma matsudairae* SV
L: 24. W: 56. All-dark storm petrel, larger than but very similar to Swinhoe's Storm Petrel, but separated by *white patch at the base of the outer primaries on the upperwing*. This patch can be visible from quite a distance, but may be weak or even lacking in some individuals, making identification difficult. Swinhoe's usually shows separate, narrow, long white shafts to outer primaries, not a solid white patch. Flight similar to Swinhoe's. **Habitat** Maritime. **Note** Rare summer visitor to Oman and UAE from Japan.

White-faced Storm Petrel *Pelagodroma marina* V
L: 20. W: 45. Dark above with *white underparts, underwing-coverts, forehead and cheeks, with dark patch through eye*. Brownish-grey upperwing-coverts contrast with blackish flight feathers and *grey rump; forked tail blackish. Long legs (occasionally dangled) project distinctly beyond the tail*. Strong and erratic, banking flight; often 'bounces' off sea in 'yo-yo' fashion. **Habitat** Maritime. **Note** Vagrant Oman, Yemen; probably regular in Arabian Sea in summer, presumably from Australasia.

Black-bellied Storm Petrel *Fregetta tropica* V
L: 20. W: 45. Small, square-tailed sooty petrel with white underwing-coverts and underbody, with an *inconspicuous black line from breast to vent*; feet project slightly beyond tail. Upperwing-coverts show obscure pale band. Erratic zigzagging flight; often dangles feet and uses them to 'bounce' on water. Easily confused with White-bellied Storm Petrel (not yet recorded in Oman). **Habitat** Maritime. **Note** Vagrant Oman, Yemen; probably regular in Arabian Sea in summer, from Southern Ocean.

Wilson's Storm Petrel

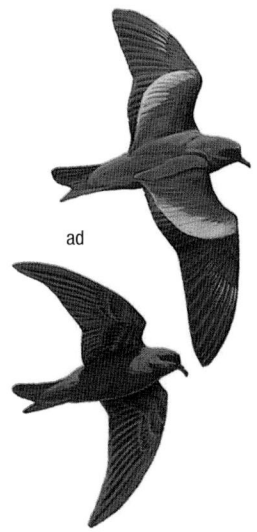
Swinhoe's Storm Petrel

Matsudaira's Storm Petrel

White-faced Storm Petrel

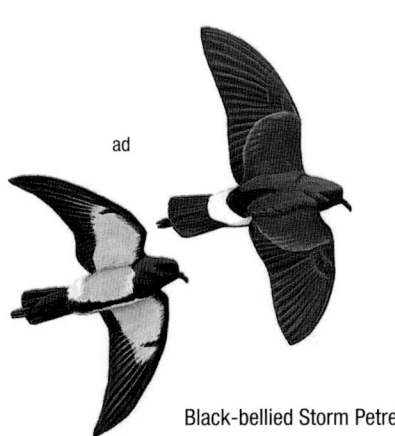
Black-bellied Storm Petrel

PLATE 8: SHEARWATERS I

Streaked Shearwater *Calonectris leucomelas* V
L: 48. W: 120. Slightly smaller than Cory's and Scopoli's Shearwaters, which it resembles in general plumage and flight, but distinguished by *white face and off-white head, streaked with brown*. Upperparts dark brown, darker than both Cory's and Scopoli's and usually lacking any pale on uppertail-coverts. Underparts white with dark patches near the shoulders, more restricted and contrasting than in Cory's and Scopoli's. Underwing white with dark flight feathers but with *diagnostic dark patch on primary coverts*. Bill slender, grey (never yellow like Cory's and Scopoli's) with darker tip. Flight similar to Cory's and Scopoli's but with slower, more elastic wingbeats and less shearing. **Habitat** Maritime. **Note** Vagrant Gulf of Aqaba and Arabian Sea, from western Pacific.

Scopoli's Shearwater *Calonectris diomedea* V
L: 46. W: 113. Large shearwater with *yellowish bill and lack of clear demarcation between grey-brown head and white chin and throat*. Underwing white with narrow brownish margins and may show ill-defined whitish band at base of uppertail. Flight relaxed with fairly slow wingbeats interspersed with long glides, often in arcs above the waves, on *characteristic bowed wings*. Scopoli's Shearwater has been recently split from Cory's Shearwater and they remain very difficult to separate. *Scopoli's is slightly smaller than Cory's, with thinner, duller bill, slightly narrower wings with white on underwing typically extending in fingers on the primaries*, but separation obviously only possible at close range or by careful examination of photographs taken. **Habitat** Maritime. **Note** Vagrant Iran and Oman, from Mediterranean (all older records assumed to be Scopoli's and not Cory's Shearwaters).

Cory's Shearwater *Calonectris borealis* V
L: 46. W: 113. Large shearwater almost identical to Scopoli's Shearwater (which see), but slightly *heavier with brighter yellow bill and more solidly demarcated dark ends to primaries below*. Most will be indistinguishable unless seen well at very close range. **Habitat** Maritime. **Note** Vagrant Oman and UAE, from Atlantic Ocean.

Persian Shearwater *Puffinus persicus* SV
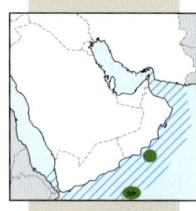
L: 32. W: 70. Small; upperparts brownish (but looks black unless close) extending to about level of eye, often showing *suggestion of a whitish forehead and supercilium, especially in autumn birds*. Underwing with *dark axillaries extending in a dark wedge onto most of the secondary coverts and a broad dark trailing edge to whole wing. Legs pink*. Fast wingbeats interspersed with short low glides, though often rising in low arcs in strong winds. At fish shoals patters over sea with outstretched wings. **Habitat** Maritime. **Note** Throughout year in hatched area; rare in Arabian Gulf.

PLATE 9: SHEARWATERS II AND JOUANIN'S PETREL

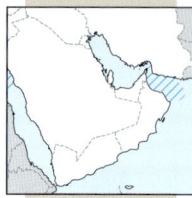

Sooty Shearwater *Ardenna grisea* V
L: 42. W: 100. Dark brown shearwater, similar in size to Wedge-tailed Shearwater but with shorter tail, much narrower wings and *variable white on underwing*, which can often flash white in contrast to dark plumage; dark feet project beyond tail in flight. Flight fast and direct with fast wingbeats interspersed with glides on stiffly held wings, sometimes arcing high above the waves. **Habitat** Maritime. **Note** Regular in summer in hatched area, though rare; vagrant The Gulf and Iran.

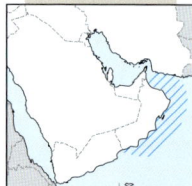

Flesh-footed Shearwater *Ardenna carneipes* V
L: 43. W: 100. Uniformly dark nut-brown, often with slightly paler coverts and *silvery flash on primaries below*, visible even at distance, distinguishes from Wedge-tailed Shearwater; if close, note Flesh-footed's larger, paler bill, slightly larger size, broader wings and *shorter tail*. Lazy flight, generally more steady than Wedge-tailed with straighter wings: a series of slow, heavy flaps followed by a long glide, rising well above the waves in windy conditions. **Habitat** Maritime. **Note** Regular in summer in hatched area; vagrant Gulf of Aqaba and UAE.

Wedge-tailed Shearwater *Ardenna pacifica* SV
L: 43. W: 100. *Sooty-brown* with longish tail, *the wedge shape being difficult to observe unless tail is spread*. May show paler bar on greater coverts above and *light edging to all coverts, forming narrow bars. Bill grey with dark tip*. Flight lazy with 3–4 fairly quick wingbeats followed by short glide, wings bowed forwards and downwards; flight erratic when windy, often changing direction and soaring in low arcs. Told from Jouanin's Petrel by larger size, *broader-based wings, longer grey bill with dark tip, which is held horizontally*, and pale feet; from Sooty Shearwater by longer tail, broader wings, *all-dark underwings and pale feet*; and from Flesh-footed Shearwater by slightly smaller size, all-dark underwing (pale underside to primaries in Flesh-footed) and thinner bill, which is grey with dark tip (flesh with dark tip in Flesh-footed). **Habitat** Maritime. **Note** Rare in summer in Arabian Sea; vagrant UAE.

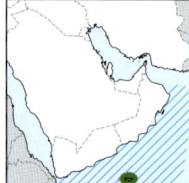

Jouanin's Petrel *Bulweria fallax* SV, sb?
L: 30. W: 75. Smaller than similar sooty-brown Wedge-tailed Shearwater with less languid flight; *all-dark stout bill held down at 45° angle*. Tail fairly long, the graduation in tail not visible unless close. In windy conditions will rise 2–5m above the waves in *long banking arcs, interspersed with spells of 4 or 5 rather leisurely flaps*, which tend to be at the peak of the arcs. Otherwise flies fairly close to the sea, progressing through troughs with a mixture of wingbeats and long glides. **Habitat** Maritime. **Note** Regular in hatched area throughout year; vagrant UAE. May breed Oman.

PLATE 10: GREBES AND FLAMINGOS

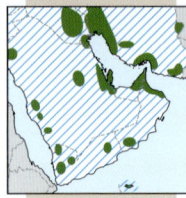

Little Grebe *Tachybaptus ruficollis* RB, wv
L: 27. W: 43. Small size, *blunt-ended body*, short neck and bill. *In flight, amount of white on secondaries and inner primaries varies, being obvious in the subspecies capensis* (which occurs in Arabia). Adult has *bright chestnut throat and cheeks, and conspicuous yellow gape-patch*. In winter, adult and young paler brown above, buff below, variably mixed with dull chestnut on foreneck (adult); light gape often reduced or absent. Juvenile has white striped head. Dives with fast jump; when alarmed dives rather than flies. Flight is only for a short distance, with rapid wingbeats low over water. **Voice** When breeding more often heard than seen; distinctive *high-pitched trilling* recalling whinny of horse. **Habitat** Well-vegetated lakes and pools; also estuaries in winter. **Note** Passage and winter hatched.

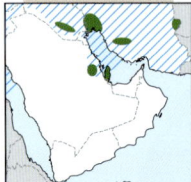

Great Crested Grebe *Podiceps cristatus* V
L: 49. W: 88. *Large grebe with long, white slender neck with pink, dagger-like bill* held horizontally. In flight, extended neck and feet are held below line of body; wingbeats rapid, *large white patch on secondaries, white border to forewing and on shoulders along body*. Easily visible on open water and usually submerges smoothly, without leaping. Breeding adult told by *black crest and black and chestnut tippets*. Winter plumage lacks tippets, but retains short, dark grey crest; *black lores and narrow white line over eye*, making eye clearly visible at distance. Juvenile has black-and-white striped head and neck. **Voice** In breeding season, a loud, harsh, far-carrying *rah-rah-rah....* **Habitat** Open freshwater, also coastal in winter. **Note** Winter hatched; vagrant Oman, UAE.

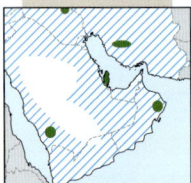

Black-necked Grebe *Podiceps nigricollis* WV, rb
L: 31. W: 58. Small grebe with *short, uptilted bill, steep forecrown and peaked head*; sometimes looks puff-backed, recalling Little Grebe. Breeding plumage *mostly black (including foreneck)* with golden ear-coverts and chestnut flanks. Winter plumage basically black and white. Uptilted bill can be difficult to see at distance. In flight long, white wing-patch extending to inner primaries; also lacks white on inner forewing. **Voice** In breeding area a plaintive whistle *ooo-eep*. **Habitat** Freshwater lakes and pools, also coastal in winter. **Note** Has bred Oman; passage and winter hatched.

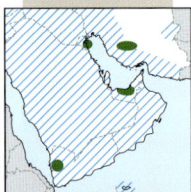

Greater Flamingo *Phoenicopterus roseus* WV
L: 130. W: 155. Very large, long-legged, long-necked, with characteristic bill shape. The white plumage of adults gradually acquires a pink hue and red wing-coverts. Juveniles, which are half the height of adults, are greyish with brown markings. Most readily told from Lesser Flamingo by larger size (but beware, as there is much size variation in Greater) and *pink (adult) or greyish (juvenile) bill with black tip* (all blackish-red bill in Lesser Flamingo). **Voice** Cacophony of deep honks and grunts, strongly recalling domestic geese. **Habitat** Coastal lagoons, salt-lakes, mudflats; breeds colonially on mud banks or in shallow water of salt lakes, building mud-heap nest. **Note** Passage and winter hatched.

Lesser Flamingo *Phoeniconaias minor* V
L: 85. W: 130. *Smaller than Greater Flamingo and generally deeper and brighter pink with all-blackish-red bill*, though very close views show dark carmine near tip. In full adult plumage has deep rose-pink on face bordering bill (lacking in Greater) and on the long scapulars (pinkish-white to white in Greater). In flight, shows rose-pink patch across centre of upperwing-coverts, bordered pinkish-white (in Greater all secondary coverts are rose-pink, but can bleach to white). Iris red (pale yellow in Greater). *Juvenile similar to Greater Flamingo in plumage but blackish bill readily identifies it*. Lesser Flamingo's smaller size can often be difficult to establish in lone birds; also, immatures of both species are smaller than adults. *In flight, note shorter neck and legs, faster wingbeats and more uptilted chin* than Greater Flamingo. **Voice** Goose or shelduck-like but wavering (yodeling); also thin high-pitched notes recalling large gulls on a rubbish tip. **Habitat** As Greater Flamingo. **Note** Regular Yemen; vagrant Iran, Kuwait, Oman, UAE.

ad winter

Little Grebe

ad summer   ad winter

ad winter

Great Crested Grebe

ad summer

ad summer

ad winter   Black-necked Grebe

Lesser Flamingo

Greater Flamingo

ad

ad

juv

juv

PLATE 11: STORKS AND CRANES

Black Stork *Ciconia nigra* wv
L: 95. W: 150. Glossy-black stork with white lower underparts. Told by *all-black upperparts* (no white on lower back and rump, as in Abdim's Stork) and *small white axillary patch on black underwing*. Adult has red bill and legs, whereas browner, less glossy juvenile has greyish-green bill and legs. **Habitat** Lakesides, marshes, riversides and fields on migration. **Note** Passage hatched; occasional in winter in S Arabia including S Oman; vagrant Bahrain, Kuwait, Qatar, UAE.

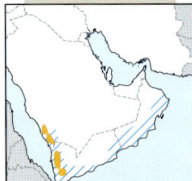

Abdim's Stork *Ciconia abdimii* wv
L: 80. W: 140. Very similar to Black Stork but smaller with *white on lower back and rump noticeable in flight*, when also shows more extensive white on underwing-coverts, less protruding greenish legs and shorter neck. At rest often shows white extension above bend of wing; bill, which is shorter than Black Stork, greyish-green, tipped reddish. Note crimson surround to eye, a small white forehead spot and bluish bare cheeks. Juvenile browner, lacking the greenish-purple gloss of adult; bill dirty flesh and bare skin of cheeks whitish-blue. Often in flocks; circles or glides at height. **Habitat** Dry plains and foothills; nests singly or colonially in trees, on rooftops or pylons. **Note** Throughout year in hatched areas, but regular in Oman in winter only.

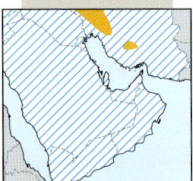

White Stork *Ciconia ciconia* WV, PM
L: 100. W: 170. Easily told by *large size, white plumage with black flight feathers, straight red bill and long red legs*. In flight, the neck is extended and legs protrude beyond the tail. Juvenile has duller white plumage and duller red bill and legs. From adult Yellow-billed Stork by straight red bill and all-white tail (black in Yellow-billed). **Voice** Clatters bill at nest; otherwise silent. **Habitat** Wetlands, plains and farmland; nests on buildings and trees. **Note** Passage hatched; many overwinter in S Oman.

Yellow-billed Stork *Mycteria ibis* E/I
L: 100. W: 160. Easily told in adult plumage by *slightly drooping, orange-yellow bill, bare red face and long, orange-red legs*. Plumage resembles White Stork but *tail black* (visible in flight) and *mantle and tips of wing-coverts tinged pink*. In subadult plumage sandy-buff with *some pinkish on underwing-coverts*, much duller bill and legs and greyish to pale orange facial skin. In first winter shows greyish wash, brownish underwing-coverts, yellowish-grey bill and greyish-brown legs. **Habitat** Wetlands. **Note** Vagrant Israel, Jordan, Turkey, from Africa; escaped birds in The Gulf, Oman.

Demoiselle Crane *Grus virgo* pm, wv
L: 95. W: 175. Smaller with shorter neck and bill than Common Crane, but size deceptive without comparison. Pale grey with largely black head and neck with elongated breast feathers hanging down in a narrow black fringe. Juvenile grey on head and neck with short whitish band behind eye. Immature gradually acquires adult plumage, general coloration browner with duller black parts. **Voice** Call higher-pitched than Common Crane. **Habitat** Open plains near water, arable land, wetlands. **Note** Passage hatched; rare Oman; vagrant Iran, Iraq, Kuwait, UAE.

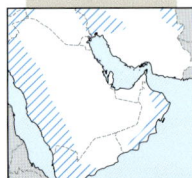

Common Crane *Grus grus* pm, wv
L: 115. W: 233. Large and majestic. *Grey plumage with contrasting black flight feathers, black head and upper neck, and white stripe from eye down side of neck*. Looks 'bushy' at rear end on ground. Juvenile has brownish head without contrasting head pattern. Adult told from Demoiselle by size and *absence of black breast*. Gregarious on migration. Neck extended in flight, as in other cranes; powerful wingbeats interspersed with long glides, often soars; flies in 'V' formation. **Voice** Often detected by far-carrying, trumpeting *krrllaa, krrllaa*. **Habitat** Wetlands, fields and steppe. **Note** Passage and winter hatched, but sporadic/local in winter; vagrant Bahrain, Kuwait, Qatar, UAE.

PLATE 12: IBISES AND SPOONBILLS

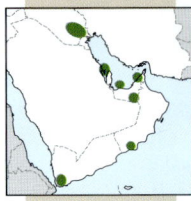

African Sacred Ibis *Threskiornis aethiopicus* V
L: 80. W: 112. Unmistakable with black-and-white plumage, *long black scapulars drooping over rear end and long, decurved bill*. Juvenile has mottled head and neck. In rather heavy flight, *shows diagnostic black line on rear edge of wings*; longish neck and legs protruding just beyond tail. **Habitat** Wetlands, cultivated areas, coastal marshes, parks, large gardens; nests colonially in trees. **Note** May breed Iran; feral population in The Gulf; vagrant Kuwait, Oman, Saudi Arabia.

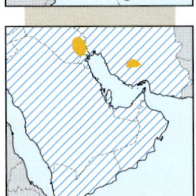

Glossy Ibis *Plegadis falcinellus* WV, pm, sv
L: 65. W: 90. *Blackish with long, decurved bill* but close views show adult to have *deep purple-chestnut plumage*, glossed green on wings. In breeding season has white marks at base of dull pink bill; in winter, bill brownish with fine pale streaks on head and neck. Juvenile much duller. Fast wingbeats in flight often interspersed with long glides; frequently flies in line-formation. **Habitat** Freshwater wetlands and marshes; nests colonially in reedbeds, occasionally in trees. **Note** Passage hatched; some winter.

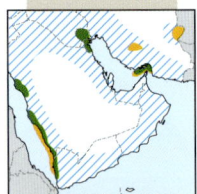

Eurasian Spoonbill *Platalea leucorodia* WV, pm, sv
L: 85. W: 120. Heron-sized with *all ivory-white plumage, characteristic black spatulate bill with yellow tip and black legs*. Adult has nape plumes and yellowish neck-band, which are lost in winter. Immature has dull flesh-coloured bill and legs and black wing-tips. In flight, neck extended and wingbeats fast and shallow, interspersed with short glides, usually in small groups in line-formation. When feeding keeps in close groups, unlike egrets and herons, sweeping head from side to side, with bill submerged. Often asleep by day, with head on back and bill hidden, but ivory-coloured plumage and closeness of birds to each other aids identification. **Habitat** Shallow open water and mudflats; nests colonially in reedbeds, islands and trees. **Note** Passage hatched; winters widely from S Turkey southwards.

African Spoonbill *Platalea alba* V
L: 90. W: 130. Similar to Eurasian Spoonbill in structure and plumage but differs in having *bare skin of face red and, in adult, grey bill, edged red, and pinkish or red legs. No nape plumes or yellow on neck*. Juvenile has reduced red skin on face (no red skin on face of Eurasian Spoonbill), black wing-tips (like young Eurasian Spoonbill) and dusky-yellow bill, which overlaps in colour with that of Eurasian; *legs of juvenile are blackish* (becoming red with age) whereas legs of young Eurasian Spoonbill dull flesh-coloured. **Habitat** Lakes, marshes, coastal lagoons and mudflats. **Note** Occasional Yemen, where may have bred; vagrant Oman.

PLATE 13: BITTERNS

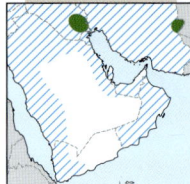

Eurasian Bittern *Botaurus stellaris* wv
L: 75. W: 130. Cryptically patterned brown heron, more often heard than seen. Smaller than Grey Heron with *stocky neck and entirely dark brown and golden-buff streaked plumage*. In owl-like flight (but with quick wingbeats) neck often extended in front, before being retracted. Juvenile Black-crowned Night Heron is smaller and shows white spots on wings; juvenile Purple Heron lacks streaks on back and has a long, thin neck. **Voice** In breeding season a *loud boom, like blowing across the mouth of an empty bottle, upwhoom*, usually repeated many times; flight call a hoarse *kaau*. **Habitat** Reedbeds; also wetlands with lush vegetation outside breeding season. **Note** Passage and winter hatched, but rare in Arabia, including Oman.

Little Bittern *Ixobrychus minutus* pm
L: 35. W: 55. *Small; conspicuous pale covert-panels contrasting with black flight feathers and dark back diagnostic*. Female Little Bittern resembles male but duller with a more rufous tinge and buff streaks on upperparts and brownish streaks on underparts. Juvenile more boldly streaked buff and brown with duller, streaked covert-panels. Legs greenish or yellowish. Most often seen in flight, when flushed, usually flying a short distance on rather jerky wingbeats before diving into cover. **Voice** In flight a short *kek* or repeated *kek-kek-kek-kek*; often calls at dusk. Male in breeding season has loud croaking *khok* repeated at intervals of about two seconds. **Habitat** Well vegetated rivers, ponds and lakes; often nests in loose colonies. **Note** Breeds S Oman; passage hatched.

Yellow Bittern *Ixobrychus sinensis* rb, mb
L: 38. Secretive; structurally similar to Little Bittern but slightly larger with *noticeably longer bill*; adult with *upperparts pinkish-brown* (not black or dark-streaked as in Little Bittern) and upperwing-covert panel slightly buffier. Juvenile and immature similar to young Little Bittern but *colours muted, less brown, the streaking being soft cinnamon-rufous with pastel-grey feather fringes, especially on underparts; mantle and back less densely streaked than Little Bittern*; crown with diffuse streaking (usually all dark in Little Bittern). Legs yellowish-green. **Voice** In flight a staccato *kakak*. **Habitat** Reedbeds, well vegetated wetland fringes, often with trees or other cover. **Note** Possibly resident Oman, common in S Oman, rare in N Oman; may breed Socotra, where also appears to be resident.

Cinnamon Bittern *Ixobrychus sinensis* v
L: 38. Adult cinnamon-tan above with *warm cinnamon ear-coverts*; in flight *wings and tail uniformly cinnamon above* (Little Bittern shows strong contrast between coverts and flight feathers in both sexes). Darker-crowned female more streaked below than male. Legs yellowish. Immature dark brown, mottled paler above; more darkly streaked below than similar-aged Little Bittern. **Voice** May give croaked *kok* when flushed. **Habitat** Wetlands, dykes and pond margins. **Note** Vagrant Oman, UAE.

Dwarf Bittern *Ixobrychus sturmii* v
L: 32. Small dark African species. Adult is dark grey above and whitish below with heavy black stripes. Immature is very dark above and dark buff below with rufous spots and stripes on wings and some black stripes on breast and belly. **Voice** Probably silent while in Oman. **Habitat** Wetlands and pool with bushes. **Note** Vagrant Oman.

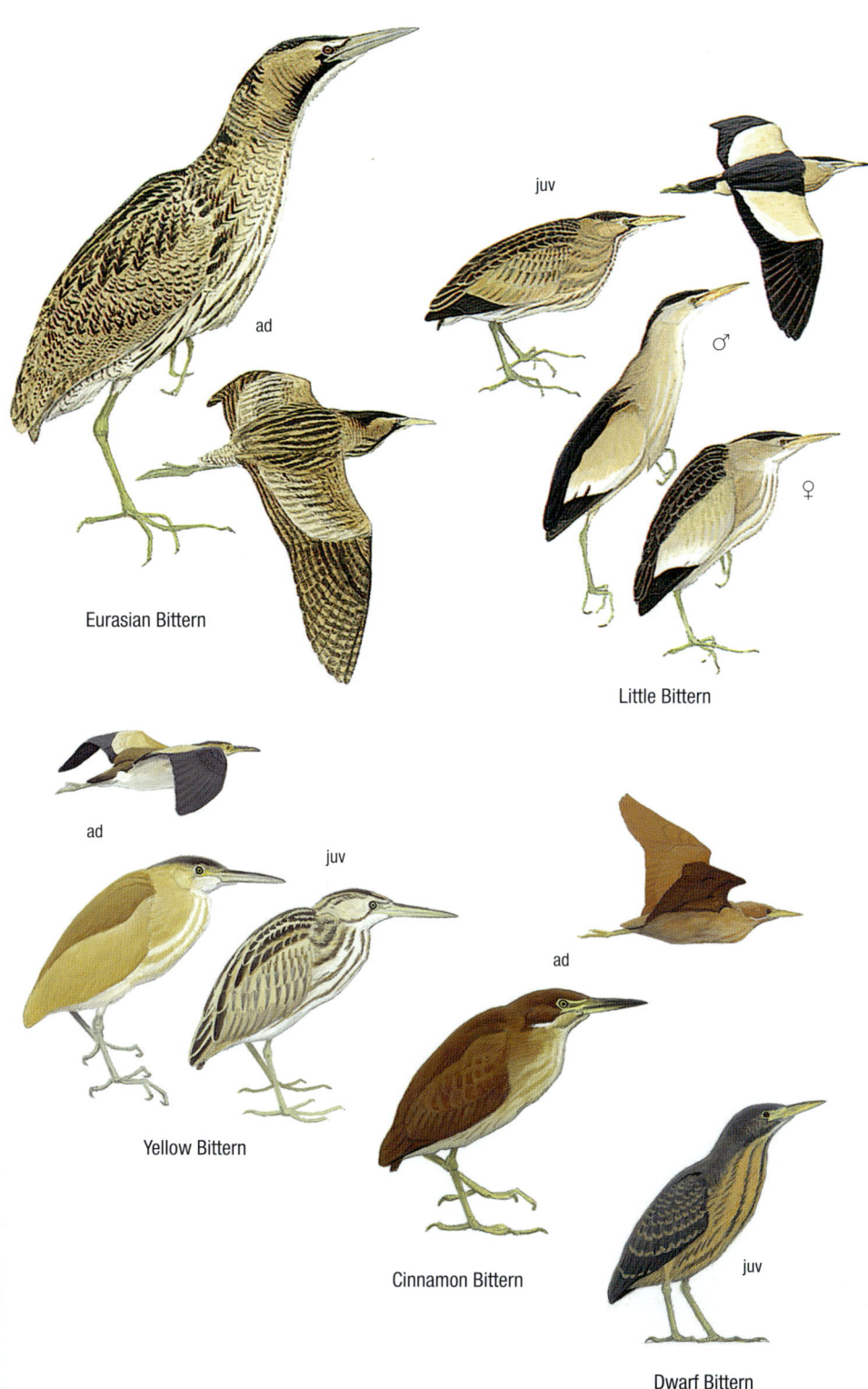

PLATE 14: HERONS I

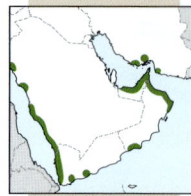

Striated Heron *Butorides striata*   RB
L: 43. W: 60. Primarily *coastal; small, rather dark heron*. Adult identified by black crown with elongated nape plumes, *bluish-grey upperparts, buff-fringed coverts and greyish neck and underparts*; yellow patch on lores and dark moustachial streak give marked facial pattern. *Rosy, pinkish or yellowish legs* extend just beyond tail in flight. Race *brevipes* may look all dark. *Immature brownish with white spots on tips of wing-coverts*, brown-and-white streaked upper breast and yellowish-green legs. May recall young Black-crowned Night Heron, but smaller with dark crown and lacks white spots on mantle. In flight (low with fast wingbeats), all ages show dark upperwings. Solitary, often skulking, adopting crouching position if disturbed; most active at dusk. **Voice** Alarm call *chook-chook-chook*; when flushed a croaking *kweuw*. **Habitat** Rocky or sandy coasts, mangroves, sometimes on inland wetlands. **Note** Range expanding in The Gulf. Subspecies in Oman not determined, but likely to include *brevipes* and *atricapilla*.

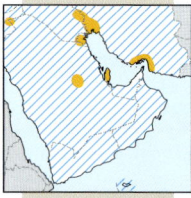

Squacco Heron *Ardeola ralloides*   WV, PM, sv, mb
L: 45. W: 85. Small heron, which usually remains in or near lush cover. Adult in breeding plumage is golden-buff with purple sheen on mantle and *long, black-and-white streaked nape plumes*; bill has a greenish-blue base in summer; nape plumes lost in winter, when neck becomes streaked and bill has yellowish base. Juvenile brownish-buff with streaked neck and upper breast, making it well camouflaged. *At all ages, white wings revealed in flight, making the bird look predominantly white. Closely resembles Indian Pond Heron (which see)*; care needed to separate in areas where both occur. **Voice** In flight, a harsh *kar*. **Habitat** Wetlands, ditches, lakesides, rivers with vegetation; nests colonially in trees, reeds. **Note** Has bred S Oman; passage hatched.

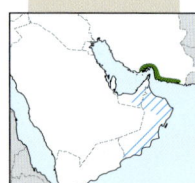

Indian Pond Heron *Ardeola grayii*   WV
L: 45. W: 85. *Very similar to Squacco Heron; often feeds in open*. Readily separable in breeding plumage by dark maroon back, unstreaked yellowish-buff to buff-brown crown and hindneck, *white nape plumes* (crown and hindneck streaked blackish, white nape plumes black-edged in Squacco), and pale buff breast (pale ochre in Squacco). In winter, loses head plumes and others are reduced in length; head, neck and upper breast become streaked brownish. *Winter adult and juvenile told from Squacco by darker, vinous-brown mantle and lack of buff tones*; dark loral line is an additional feature (though sometimes absent), the bare skin above it being rectangular in shape (but triangular in Squacco, which lacks dark loral line). Juvenile similar to winter adult but primaries dark-tipped. Legs greenish/yellowish (red in some breeding birds). **Voice** Similar to Squacco. **Habitat** Fresh and saltwater marshes especially with dense vegetation; mangrove mudflats, rivers, ponds. **Note** Some present throughout year in hatched area; vagrant Kuwait.

Chinese Pond Heron *Ardeola bacchus*   V
L: 45. W: 85. Adult in summer has *chestnut head and breast*. Dark grey back usually covers the wings. White wings and tail conspicuous in flight like other pond herons. Immature Bittern-like on head and breast, brown back. **Voice** Call a deep *qwa*. **Habitat** Wetlands and streams. **Note** Vagrant Oman.

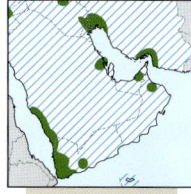

Western Cattle Egret *Bubulcus ibis*   WV, PM
L: 50. W: 85. Small, white, grassland-dwelling heron. Separated from Little Egret by stockier build, *shorter yellow bill, shorter neck and legs, extended 'jowl' under bill* and faster flight. Also in *breeding season, orange-buff wash on crown, back and breast, and reddish bill*. Usually seen feeding in small groups, associating with grazing animals, or flying in flocks to and from roost sites. **Voice** A short *ark* and duck-like *og-ag-ag*, often in flight. **Habitat** Fields, wasteland, marshes; nests colonially in reedbeds, bushes or trees. **Note** Passage and winter hatched.

Eastern Cattle Egret *Bubulcus coromandus*   V
L: 50. W: 85. Very similar to Western Cattle Egret, but has *longer bill and tarsi*. In summer plumage *buff on head reaches cheeks and throat*. In winter plumage similar to Western Cattle Egret and doubtless overlooked. **Habitat** Fields, wasteland, marshes. **Note** Vagrant Oman, UAE.

PLATE 15: HERONS II

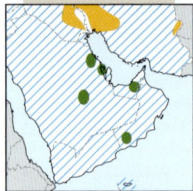

Black-crowned Night Heron *Nycticorax nycticorax*   PM, WV
L: 60. W: 110. Stocky, about half the size of Grey Heron; most active at dusk. *Adult's grey plumage with black back and crown is unmistakable. Brownish juvenile is prominently spotted with white on back and coverts* and this feature immediately identifies it from other brown herons. By second calendar year the spotting is reduced in size and the mantle and scapulars become grey-brown (mirroring the black of the adult). More likely to be flushed than seen in the open during the day. **Voice** Harsh deep *kwark*, heard in flight, especially at dusk. **Habitat** Rivers, lakes and marshes with trees and dense vegetation; nests in reeds or trees. **Note** Has bred Kuwait, S Oman; passage and winter hatched, but few winter in the north.

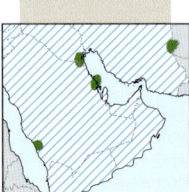

Grey Heron *Ardea cinerea*   PM, WV, sv
L: 95. W: 185. Large with long neck and legs and powerful bill. In all plumages has *grey upperparts and white underparts and neck*. Adult has *black crest and dark markings down front of neck*. Juvenile has darker grey or browner upperparts and crown with rest of plumage white. Bill yellowish and legs brownish-yellow in adult, but in younger plumages bill browner and legs greyish. Juvenile can superficially resemble Purple Heron, but is much paler with thicker neck, stouter bill and never shows buffish-brown on neck and breast. **Voice** A raucous *waak* in flight. **Habitat** Wetlands including coasts; nests in trees. **Note** Passage and winter hatched; many oversummer in Oman and UAE.

Black-headed Heron *Ardea melanocephala*   V
L: 85. W: 160. Slightly smaller than Grey Heron with *black crown and hindneck contrasting with white throat and foreneck* in adult plumage. Grey upperparts, darker than Grey Heron, and legs blackish; bill has grey upper mandible (yellowish in Grey Heron). *At all ages has white underwing-coverts contrasting with black flight feathers*. In juvenile, crown and hindneck dark grey, contrasting less with whitish foreneck, which is often tinged rusty. **Habitat** Inland or coastal wetlands. **Note** Regular throughout year in Yemen, where may breed; vagrant Oman.

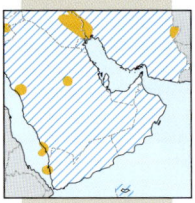

Purple Heron *Ardea purpurea*   PM, WV
L: 80. W: 135. Slimmer and much darker than Grey Heron, with *angular, snake-like neck*. Adult dark grey with black belly, *chestnut on neck edged by black streak, black crown and line across cheeks*; dull yellow bill. Juvenile has sandy-brown upperparts and hindneck with diffuse dark streaking on neck and blackish crown. *In flight, has more angular neck (bulging downwards more obviously) than heavier-looking Grey Heron with prominent spread toes and brownish upperwing-coverts*. Mostly seen in or when flushed from vegetation, unlike Grey Heron which feeds in more open areas. **Habitat** Marshes, reedbeds, ditches; nests colonially in reeds or trees. **Note** Mainly summer visitor; passage hatched.

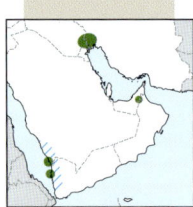

Goliath Heron *Ardea goliath*   V
L: 145. W: 220. *Very large; with long legs and heavy bill*. Adult has chestnut head and hindneck, bluish-grey upperparts and *rich chestnut underparts and underwing-coverts*. Immature birds are *paler rufous-orange on head* and have rufous edgings to feathers on upperparts; white underparts are dark-streaked. *Told from Purple Heron by much larger size, stouter head and neck, lack of black on crown, absence of black line across cheeks, greyish bill* (yellowish in Purple), *dark legs* (yellowish-green in Purple) and upperwing more uniform grey. In flight, wingbeats slow and heavy with legs protruding well beyond tail. **Habitat** Extensive reedbeds; also mudflats and coastal islands; nests in mangroves and reeds. **Note** Feral breeding UAE; vagrant Oman.

PLATE 16: HERONS III

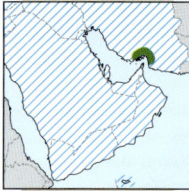

Great Egret *Ardea alba*   WV, PM
L: 95. W: 155. *Largest white heron with long, angular neck often stretched to full extent.* In breeding plumage note scapular plumes, black bill with yellow base and black legs, yellowish above the joint. Winter adult and juvenile have yellow bill and blackish-green or brownish legs. *Told in all plumages from Intermediate Egret by longer, thinner bill and black gape-line extending behind eye.* Sedate in flight and on ground. **Habitat** Wetlands and rivers. Nests colonially in reeds or trees. **Note** Passage and winter hatched. Smaller eastern subspecies *modestus* is rare breeder in Iran, and occasional visitor to SE Arabia. [Alt: Great White Egret, Western Great Egret]

Intermediate Egret *Egretta intermedia*   pm, wv
L: 65. W: 110. All white; between Great Egret and Little Egret in size. Bill orange-yellow; in non-breeding season usually has black bill tip in Asian subspecies, *intermedia*, which is increasingly common, especially in S. Oman (though bill all-black in breeding season). Yellowish facial skin and blackish legs with brownish-grey joints and tibia. Most easily confused with Great Egret but note smaller size, shorter neck and *shorter, stouter bill*. *Best told by very short gape-line, which does not extend behind eye (as it does in Great Egret).* Can appear similar to Little Egret and white morph of Western Reef Heron but larger, with longer legs and blackish feet; bill also brighter yellow and straight. **Habitat** Wetlands, coastal and inland. **Note** Passage and winter hatched; vagrant UAE, Yemen.

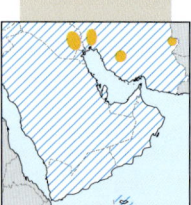

Little Egret *Egretta garzetta*   WV, PM
L: 60. W: 90. Graceful, clean white egret which, in adult plumage, has *all-black bill and black legs with yellow feet*. In breeding season shows long, delicate plumes on nape and mantle. Juvenile has brownish-green legs and pinkish base to lower mandible. May be confused with white morph of Western Reef Heron, which see. Told from cattle egrets by larger size, bill colour and lack of buffish wash to plumage. **Voice** Throaty grunting *raaak* in flight. **Habitat** Wetlands; nests colonially in trees and reedbeds. **Note** Passage and winter hatched.

Western Reef Heron *Egretta gularis*   RB, WV
L: 60. W: 90. White, dark and intermediate morphs occur; subspecies *schistacea* (Indian Reef Heron) occurs. White morph similar to Little Egret but less elegant, *thicker-billed with curved culmen giving slightly drooping look*. Bill in all plumages is pale brown to yellowish, usually with reddish flush in breeding season. Legs dark olive-brown, but tarsus often greyer-greenish up rear, sometimes also on lower forepart or even entirely greenish, with the feet generally greenish or yellow (as Little Egret), very rarely rosy. Outside breeding season facial skin yellow or greenish-yellow (blue-grey in Little Egret). Dark morph *slate-grey with white chin and throat* and occasionally a few white flight feathers; juvenile dark morph is paler grey than adult with whitish on foreneck, breast and belly; like adult can show white feathers in wing. Juvenile white morph often has grey feathers in plumage. See also Intermediate Egret. Feeds from rocks or by slowly wading in shallow water, along tide edge or beach with occasional sudden dash after prey, often with wings flailing. **Voice** Guttural *grrurr*. **Habitat** Coastal, especially tidal flats; rarely inland; nests colonially in mangroves, low bushes on offshore islands. **Note** Local breeder in Oman; common on passage and in winter. [Alt: Western Reef Egret]

Black Heron *Egretta ardesiaca*   V
L: 50. W: 70. *A small, all-black heron*. Short, ragged crest, black bill and legs with orange-yellow feet. Juvenile blackish-brown and lacks crest. Very active when feeding, often throwing wings over head, creating a shading umbrella canopy over the water. **Habitat** Wetlands, marshy fringes, tidal flats. **Note** Vagrant Oman, Yemen.

PLATE 17: PELICANS, TROPICBIRD AND FRIGATEBIRDS

Great White Pelican *Pelecanus onocrotalus* wv
L: 140–175. W: 270–330. Large with huge wingspan. Adult *white with contrasting, solid black flight feathers below*, but body tinged yellowish-rosy in breeding plumage (Dalmatian Pelican has greyish underwing, and body appears greyish-white). At close range note *short, shaggy crest on nape* (in breeding season), *dark eye surrounded by naked rosy skin, fleshy-yellow legs and pointed forehead feathers where they meet the culmen*; these latter characters are also useful when separating immature from similar Dalmatian Pelican. Immature Great White Pelican has *clearly darker grey-brown upperparts than the grey-buff Dalmatian*. Flight consists of a few slow wingbeats followed by a glide; flocks often fly in regular lines, or circle in formation. **Habitat** Large inland wetlands and shallow coastal lagoons; nests colonially in reeds. **Note** Has bred Kuwait; passage hatched, some winter S Turkey southwards; rare Oman; vagrant Bahrain, UAE.

Pink-backed Pelican *Pelecanus rufescens* v
L: 130. W: 265–290. Smaller than Great White Pelican (hard to judge without comparison). *Adult duller and greyer than other pelicans with darkish curly crest on nape. In flight, shows dull greyish flight feathers below* (bleaching paler), *separated from whitish-grey or pinkish-rufous underwing-coverts by whitish translucent band through centre of wing* (pattern similar to Dalmatian Pelican). *Flanks, back and rump have pink tinge in breeding season*. At close range told from Dalmatian Pelican *by black markings around dark eye and pale legs*. Immature brownish above and on tail, whitish below; in flight broad white rump narrows into white band up centre of mantle, almost to base of neck, framed by brownish shoulders and wing-coverts. Never shows solid black flight feathers below as in Great White Pelican. **Habitat** Coastal waters and shores; nests in mangroves and on sandy islands. **Note** Some dispersal as hatched; vagrant Oman.

Dalmatian Pelican *Pelecanus crispus* v
L: 150. W: 310–345. Resembles Great White Pelican but told in flight by *greyish underwing with pale band through centre and greyish-white body* (in Great White, flight feathers below solidly black, and white body tinged with yellowish-rosy). At close range note *nape feathers curl upwards* (drooping in Great White), *pale eye* (dark in Great White) *and grey legs*; also shape of bare skin around eye and of feathers where they meet the culmen useful at all ages. Immature dirty white below, *pale grey-buff above* (similar Great White is dark grey-brown above). **Habitat** As Great White Pelican; nests in reeds and trees. **Note** Partial migrant; vagrant Kuwait, Oman, UAE.

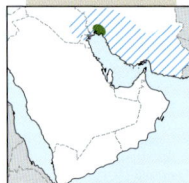

Red-billed Tropicbird *Phaethon aethereus* MB
L: 48 (plus 50cm tail). W: 105. Unlikely to be confused with any other seabird in the Middle East. Plump-bodied, white with exceptionally *long, white tail-streamers and conspicuous red bill*; the white plumage is relieved by a black eye-stripe, black outer primaries and narrow black barring on upperparts and coverts. Juvenile (which lacks tail-streamers), has black-tipped tail, yellowish bill and blackish collar. Flight is a useful character: direct with *fast wingbeats and interspersed with glides on horizontally held wings*, usually fairly high. Will settle on sea. **Voice** Shrill, rapid rasping notes. **Habitat** Maritime; nests colonially on rocky mainland or island cliffs, or rocky slopes on islands. **Note** Occurs at sea in hatched area throughout year; vagrant Kuwait.

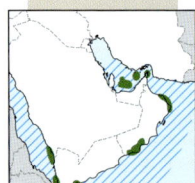

Great Frigatebird *Fregata minor* v
L: 93. W: 218. Large, dark, piratical seabird with long, narrow wings and long, deeply forked tail (often held closed). Very difficult to separate from Lesser Frigatebird but note *absence of white on axillaries, at all ages; adult male is all black* and thus most easily distinguished. See Lesser Frigatebird for further differences. Soars and glides majestically, with only an occasional deep wingbeat. Does not settle on sea. **Habitat** Maritime but may drift inland or be driven onshore by tropical storms. **Note** Vagrant Oman, Socotra.

Lesser Frigatebird *Fregata ariel* v
L: 75. W: 185. Smaller than but otherwise very similar in structure and flight to Great Frigatebird, and unless good views of the underwing are obtained most frigatebirds in the field cannot be identified. *White extending onto axillaries is the most useful field feature at all ages*. Female has black head, which when seen from below contrasts with white on breast (white chin, throat and breast in Great Frigatebird). Juvenile often shows black mottling on lower white breast whereas in Great Frigatebird the white breast-patch is neater. Does not settle on sea. **Habitat** Maritime but may drift inland or be driven onshore by tropical storms. **Note** Vagrant Gulf of Aqaba, Oman, Yemen.

PLATE 18: GANNET, BOOBIES AND CORMORANTS

Cape Gannet *Morus capensis* V
L: 85. W: 150. Adult very similar to extralimital Northern Gannet; slightly smaller with *all-black secondaries, above and below,* and black tail. Subadult Northern Gannet can also show black secondaries and black in tail but the white underwing-coverts will always show dark smudgy markings and there will never be the clean-cut contrast between pure white coverts and black flight feathers of Cape Gannet. Juvenile similar to juvenile Northern and only identifiable if caught or found dead, when note the long, black, bare gular stripe on chin and throat (short in Northern). Flight similar to Northern Gannet but slightly faster wingbeats. **Note** Vagrant Oman, from southern Africa.

Masked Booby *Sula dactylatra* RB
L: 85. W: 150. Adult and subadult distinctive, but juvenile recalls Brown Booby. Juvenile Masked Booby has *brown head and neck separated from paler brown back by white collar,* which broadens with age, while upperparts become mottled with white especially (first) on the scapulars; underparts are white, also *underwing-coverts white with black line through centre.* Juvenile differs from Brown Booby primarily in larger size, variable white mottling on upperparts, white neck-collar and more prominent band on underwing-coverts. See also Red-footed Booby. **Habitat** Maritime. **Note** Occurs at sea in hatched area throughout year; rare in The Gulf; vagrant Iran, UAE.

Red-footed Booby *Sula sula* V
L: 75. W: 100. This vagrant to Arabian seas has a white and a brown colour morph, which can be can be similar in some plumages to Masked and Brown Boobies. Birds from the Indian Ocean (the most likely to occur) have, in *adult plumage, irrespective of morph, all-white tails. Adult white morph resembles Masked Booby but is smaller, has a white tail, black carpal-patches below,* often a yellowish wash to the head, and *red feet;* lacks a black mask. *Brown morph adult is grey-brown* with darker back and wings (both above and below) and *white tail.* Juvenile is rather featureless, all brown with noticeably dark underwing, lacking white or pale on coverts (as shown by juvenile Brown Booby). **Habitat** Maritime. **Note** Vagrant Oman, UAE, from Indian Ocean.

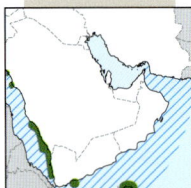

Brown Booby *Sula leucogaster* WV, SV
L: 70. W: 145. The commonest booby in the Red and Arabian Seas and readily identified in adult plumage by *uniform chocolate-brown upperparts, head and neck, and conspicuously white underbody and underwing-coverts.* The *pale greenish-yellow bill* contrasts with the dark head even at a distance. Juvenile plumage similar to adult but underparts buffish-brown (thus less contrast with brown head, neck and upper breast); white underwing-coverts appear at an early age but in some very young birds can look quite brownish. (See Masked Booby for differences.) Flight comprises a series of wingbeats followed by a long glide with the wings held fairly horizontal. Catches fish by diving, with folded wings, often at a shallow angle from a short height above the sea. See Masked Booby for separation of respective juveniles. **Habitat** Maritime. **Note** Occurs at sea in hatched area throughout the year; rare in The Gulf.

Great Cormorant *Phalacrocorax carbo* WV
L: 90. W: 140. Large; swims low in water and frequently perches with wings outstretched. Breeding birds have *white on nape and neck and white thigh-patch. In winter retains white patch on chin and throat,* unlike Socotra Cormorant. Juvenile browner than adult with dirty white underparts. **Habitat** Coasts and inland lakes. **Note** Has bred Bahrain; passage and winter in hatched area.

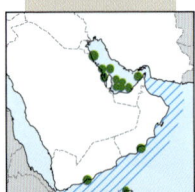

Socotra Cormorant *Phalacrocorax nigrogularis* RB, wv
L: 80. W: 130. Slightly smaller than Great Cormorant, with slimmer head and neck (resembling extralimital European Shag *P. aristotelis* in structure, though longer-winged in flight). *Adult sooty-black with glossy bronze-green wings and back, lacks white face and chin-patch of Great Cormorant and has much slimmer greyish bill.* Immature grey-brown above with pale fringes *to coverts and scapulars;* breast and belly off-white, sometimes with brownish spotting. Juvenile has less obvious pale fringes to coverts and lacks spotting on breast and belly; best separated from young Great Cormorant by structure, bill shape, pale-fringed coverts and, when present, dark spotting below. Congregates in large flocks in and out of breeding season. **Habitat** Maritime, coastal; nests in large colonies mainly on offshore islands. **Note** Occurs in hatched area outside breeding season.

PLATE 19: VULTURES

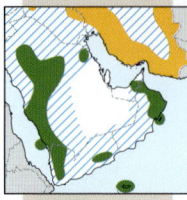

Egyptian Vulture *Neophron percnopterus*   RB, WV

L: 62. W: 155. A small vulture. Adult has *white, wedge-shaped tail*, white underparts with black flight feathers (secondaries greyish-white above), *small pointed head and thin bill*; colour pattern of plumage resembles pale morph Booted Eagle or White Stork below but shape quite different. Juvenile is mid-brown below with blackish ruff; dark brown above with creamy bars on wing-coverts, pale rump and whitish uppertail-coverts; *wedge-shaped tail grey-brown, tipped paler*. Soars on flat to slightly arched wings; active flight has many deep wingbeats between glides. Often in flocks. **Habitat** Mountains, isolated peaks, wadis and open country; frequents village refuse dumps; sometimes on the foreshore; nests on cliffs. **Note** Partial migrant in S Iran; passage hatched; vagrant Qatar.

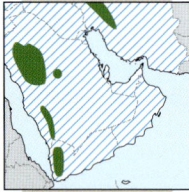

Griffon Vulture *Gyps fulvus*   wv, pm

L: 95–105. W: 245–270. Large; heavy with long, broad, deeply fingered *wings with curved trailing edge*; short, broad, square-cut tail and slightly protruding narrow head. *Soars effortlessly for long periods on raised wings*; active flight with very slow, deep wingbeats; glides on kinked wings. Adult *gingery-buff above and below contrasting with dark flight feathers*. Juvenile even paler brownish-yellow on rear underwing-coverts, thus greater contrast with flight feathers. Gregarious. **Habitat** Mountains; occurs over all types of country in search for food; nests colonially in caves or on cliff ledges. **Note** Passage, winter and dispersal areas hatched, but rare in much of Arabia. [Alt: Eurasian Griffon Vulture]

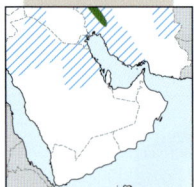

Cinereous Vulture *Aegypius monachus*   V

L: 105. W: 255–295. Very large. Readily told from Griffon Vulture by all-blackish plumage without any contrast and parallel-edged wings held flat, or slightly downcurved, particularly when gliding; tail also slightly longer and less square-cut than in Griffon. Young birds blacker than adults, but in both pale legs stand out against black undertail-coverts. At close range adult has black and whitish head pattern; head blackish-brown in juveniles. Plumage blacker throughout than rather similar Lappet-faced Vulture. Told from dark eagles by larger size, longer and more deeply fingered wings, and less protruding head. The occasional wingbeat is slow and deep (like Lappet-faced). **Habitat** Desolate mountains (often extensively wooded), foothills, plains and semi-deserts; nests in trees, sometimes on cliffs. **Note** Winter dispersal hatched; vagrant Oman. [Alt: Eurasian Black Vulture]

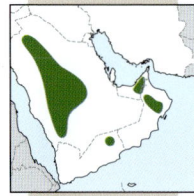

Lappet-faced Vulture *Torgos tracheliotos*   RB

L: 105. W: 255–290. Very large, heavy vulture, *paler than Cinereous Vulture but darker than Griffon*. Long, deeply fingered wings and short tail with distinctly pointed feather tips; wings less parallel-edged than in Cinereous but less curved than in Griffon. *From above, dull grey-brown wing-coverts contrast much less with flight feathers than in Griffon* (but bleach paler; in Cinereous virtually no contrast). From below, dark wing-coverts have variable whitish 'vulture streak' near leading edge and *flight feathers and their coverts are clearly paler* greyish, but wing-tip blackish (in Cinereous, black underwing-coverts contrast with paler flight feathers, and wing-tip not clearly darker). Dark brown underparts have whitish-brown mottling on breast, creamy upper flanks, browner upper thighs but paler lower thighs and ventral region, *giving underparts a variegated appearance*. Immature birds are more uniform with less developed 'vulture streak' and less variegated underparts, but usually have some pale on vent. When perched, adult identified by very heavy bill, feathered hindneck, ugly unfeathered pinkish head and foreneck, and long, lanceolated breast feathers; *lappets often inconspicuous*. Solitary or in pairs, but small parties sometimes at carcasses. **Habitat** Savanna, semi-desert steppe, desert with scattered trees, foothills, rocky wadis; huge nest built on top of an acacia. **Note** Some winter dispersal; vagrant Kuwait.

PLATE 20: OSPREY AND HONEY BUZZARDS

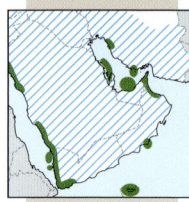

Western Osprey *Pandion haliaetus* RB, WV
L: 56–61. W: 145–165. *Large; long, narrow wings, distinctly angled when gliding, white undersurface with black carpals and band through centre of underwing, white crown and dark eyemask*. Variable dusky band across foreneck (usually boldest in female). Juvenile has whitish scales and white line on greater coverts above. Flies with steady, shallow wingbeats, glides on smoothly curved wings; may recall large soaring gull at distance. Hovers over water for fish and dives with splash, feet first, almost disappearing. **Habitat** Always near water, inland or coastal; nests in trees, on sea-cliffs, remote islands (often on ground), ruins, old wrecks, sometimes in scattered groups. **Note** Passage hatched, winters on coasts.

European Honey Buzzard *Pernis apivorus* pm
L: 55. W: 135–150. *Recalls Steppe Buzzard in shape but slimmer with longer, narrower tail with rounded corners; head and neck narrower, protruding in cuckoo-like manner. Wingbeats more flexible and soars on flattish wings, and glides on slightly lowered wings* (Steppe Buzzard soars on raised wings). Plumage variable; typical male has greyish head and upperparts; female browner. Below, some are dark, others largely white, *but most are barred on body and coverts, and have black carpal patches; flight feathers show prominent black trailing edge and characteristic bars at base (more bars in female). In all morphs, tail has dark band at tip and two bars at base*. Cere grey, eyes yellow or orange-yellow (male). Juvenile dark brown, rufous-brown or creamy-white with streaked breast; *usually, but not always, with dark carpal patches and narrow whitish crescent on uppertail-coverts; head often whitish with dark eye-mask*; may show pale band on underwing-coverts, separating secondaries from dark forewing (unlike Steppe Buzzard) and *three evenly spaced bars on flight feathers* (unlike adult). Juvenile with its more slender wings with curved rear-edge (bulging secondaries) and shorter tail has more of a Steppe Buzzard-like outline, but shape of tail and head, and soaring on flat wings important for identification. Migrates in flocks. *See similar Crested Honey Buzzard for separation* from that species. **Habitat** Woodland; widespread on passage. **Note** Passage hatched, but rare in E Arabia.

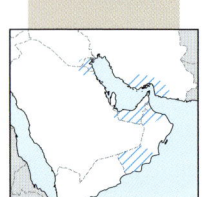

Crested Honey Buzzard *Pernis ptilorhynchus* wv
L: 65. W: 160. Resembles European Honey Buzzard but larger with noticeably broader body, slightly longer wingspan, and broader wings with bulging secondaries (all ages) and wing-tip showing *six long 'fingers'* (five in European Honey); also tail broader and shorter. Dark, pale and intermediate morphs occur (as European Honey). Adult has *dark gorget across throat and lacks contrasting black carpal patches; nape plumes sometimes visible when perched*. Male has *dark red eyes, undertail with, apart from broad black band at tail base, two black bands divided by broad pale band*, and *black band on flight feathers reaches body* (male European Honey has inner part of band hidden beneath greater coverts). Adult female undertail pattern is more like male European Honey, but innermost bar (at tip of tail-coverts) usually broader; secondaries crossed by three dark, evenly spaced bars (female European Honey usually has two bars with wider gap between dark trailing edge and first bar). Juvenile has underparts creamy to foxy or dark brown, dark eyes, yellow cere, and variable tail barring, typically four narrow bars of even width, all being characters similar to juvenile European Honey and thus best told by shape and structure, but note often has *broad pale rump*, while none shows dark carpal patches (usually obvious in European Honey, except dark individuals). Adult Crested Honey Buzzard usually migrates in autumn with 4–5 new inner primaries (adult European Honey has from 0–3 renewed). Care is required with any wintering honey buzzard, as moulting Crested Honey may be missing outer primaries and then only shows five fingers (as European Honey). **Habitat** Open woodland, parks, wooded farmland. **Note** Rare passage (any habitat) and winter hatched; uncommon Oman; vagrant Saudi Arabia, Yemen.

PLATE 21: KITES

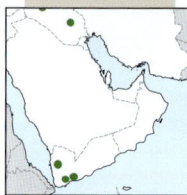

Black-winged Kite *Elanus caeruleus* pm, wv
L: 33. W: 76. Small, similar in size to Common Kestrel, but appears larger in flight; also hovers. Plumage essentially *pale grey and white, with black primaries and large black patch on forewing of pale grey upperparts*. Underparts white; eyes red. Juvenile darker above than adult with white-tipped greater coverts forming narrow line. Note well-protruding, broad head, *shortish tail* with slight notch when closed, long and pointed but relatively broad-based wings. Rather owl-like flight with soft wingbeats; wings raised, rather like a harrier when soaring; persistently hovers. **Voice** In display, a thin whistling *wee-oo, wee-oo*. **Habitat** Open country with trees. **Note** May breed Iran; rare Oman; vagrant Kuwait, Qatar, Saudi Arabia, UAE. [Alt: Black-shouldered Kite]

Brahminy Kite *Haliastur indus* V
L: 48. W: 135. Size of Black Kite, with similar wing position though will soar with wings in shallow 'V'. Adult has *white head, neck and breast*; otherwise red-brown above and on underbody with paler rufous underwing-coverts; flight feathers and undertail creamy-buff; blackish wing-tips most conspicuous from below. Juvenile lacks white head and neck, is darker brown above and below, with greyish secondaries below but *conspicuous whitish primary patch*. Sometimes recalls Marsh Harrier when foraging low over ground or water, including sea. **Habitat** Open country, coasts. **Note** Vagrant Oman.

Black Kite *Milvus migrans* PM, WV
L: 50–65. W: 125–150. Long-winged with languid, elastic wingbeats. *Tail long and forked* (often square-ended when fully spread). Adult has *hardly any white on primaries below*. Juvenile shows dark eye-mask, pale feather tips on mantle and shoulders, boldly dark-spotted breast but paler belly and diffuse dark band to tail; also *whitish tips to greater upperwing-coverts*. Soars and glides on slightly arched wings; manoeuvres tail when scanning for food. Race *lineatus* (Black-eared Kite – also illustrated) differs in having prominent white 'window' in *base of dark-barred primaries, broader hand with long sixth primary, all six 'free' primaries longer than in Black Kite, and browner head (greyish in Black Kite) with more obvious dark eye-mask*. Underbody is streaked off-white to ochre (finer pale and black streaking in Black Kite). Juvenile similar to juvenile Black Kite but has broader hand with long sixth primary (like adult), whiter base to primaries, more noticeable streaking across breast and pale belly and vent. **Voice** Gull-like, whinnying *yiieerr*. **Habitat** Woodland, often near water; anywhere on migration, often gathering at rubbish dumps. **Note** Passage hatched; birds in E Arabia probably Black-eared Kite or hybrid Black x Black-eared Kite.

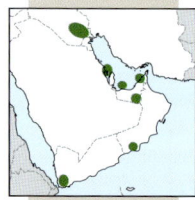

Yellow-billed Kite *Milvus aegyptius* rb
L: 47–60. W: 120–145. Adult immediately told from Black Kite and Black-eared Kite by *yellow bill*; also slightly smaller with *brighter rufous-brown underparts and more distinctly barred tail*, with slightly deeper fork. Primary pattern below similar to Black Kite, thus lacking the white primary bases of Black-eared Kite; head brownish (not greyish as in Black Kite). Juvenile sandy-brown below, streaked darker with quite prominent, lightly barred white bases to primaries – thus more like Black-eared Kite than Black Kite; bill can be all dark or yellow with black tip. **Habitat** Villages, towns, open country with trees, especially palms. **Note** Breeds S Oman (rare); vagrant Socotra.

PLATE 22: EAGLES I

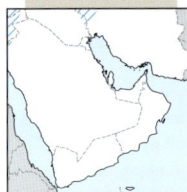

Lesser Spotted Eagle *Clanga pomarina* — pm, wv
L: 62. W: 145–165. *Medium-brown adult shows contrast between pale upperwing-coverts and darker brown mantle, underwing-coverts paler than flight feathers, a neat pale patch at base of primaries above*, and small creamy area on uppertail-coverts. Juvenile darker, warm brown below with *flight feathers of same shade or slightly darker (coverts never darker than flight feathers)*; unlike adult may show little contrast between mantle and wing-coverts; darker head has *rusty-yellow spot on nape* (absent in Greater Spotted, present in some adult Steppe Eagles); narrow white bar on greater upperwing-coverts and distinct whitish inner primary patch. *Short or minute seventh primary, less deeply fingered wings and smaller hand separates from Steppe Eagle at all ages*; lacks dark band on trailing edge of underwing and tail of many adult Steppe. Wings relatively narrow, tail medium-long; active flight less heavy than larger *Aquila* eagles. Soars and glides on arched wings with primaries lowered. On ground lacks heavy 'trousers'. *Both spotted eagles have characteristic round nostrils* (unlike Steppe). Migrates in flocks. **Habitat** Breeds in forests; open country on passage **Note** Passage hatched; rare Oman; vagrant Kuwait, Saudi Arabia, UAE, Yemen.

Greater Spotted Eagle *Clanga clanga* — PM, WV
L: 65. W: 155–180. *Typically darker than Lesser Spotted Eagle. Adult dark brown below, flight feathers similarly dark or a shade paler* (reverse in Lesser Spotted); *leading underwing-coverts sometimes blackish-brown* (never so in Lesser); on upperwing, mid- to dark brown coverts sometimes contrast with darker mantle (like typical Lesser); no conspicuous pale primary patch above (unlike Lesser and Steppe Eagle). Adult Greater Spotted usually lacks band on trailing edge of underwing, seen on many adult Steppe; also lacks pale nape-spot of young Lesser and many adult Steppe. *Juvenile is blackish-brown below with paler flight feathers* (in Lesser Spotted coverts are brown but flight feathers are never paler); *blackish-brown upperwing has 1–3 white covert bars, often creating pale panel*; large, diffuse primary patch formed by whitish primary shafts and pale inner primaries (patch smaller, more conspicuous in Lesser). Infrequently, young Greater Spotted is abnormally coloured on body and wing-coverts: i) '*fulvescens*' type – illustrated; ii) yellow-brown above and below; iii) yellow-brown above, normal below; iv) yellow-brown below, normal above; v) underwing-coverts greyish, or dark, mottled paler, underbody darker; or normal upper- and underwing, but contrasting paler underbody. Irrespective of age, *secondaries below may have thin dense bars* (broader in Lesser Spotted, more well-spaced in Steppe). Adult has relatively broad and parallel wings with slightly broader hand and deeper fingers than Lesser. Juvenile has narrower hand than adult with trailing edge fairly strongly curved inwards at body (visible when tail closed). Hand slightly shorter, less ample than Steppe (Greater Spotted has shortish seventh primary) and bill generally smaller. **Habitat** Usually near wetlands, coastal or inland, also rubbish tips. **Note** Passage and winter hatched.

Tawny Eagle *Aquila rapax* — V
L: 70. W: 165–185. Slightly smaller than Steppe Eagle. *Plumage often creamy or rufous*, unlike brownish Steppe; *creamy-white lower back and pale wedge on inner primaries below is typical*. Some adults are dark brown and hard to separate from adult Steppe, but are rare in the Middle East. Juvenile, and often adult, has pale brown plumage which bleaches to creamy-white (recalling '*fulvescens*' Greater Spotted). Adult often has small primary patch above but in juvenile patch is larger. *At all ages flight feathers below are dark to pale grey, mostly with fine, dense bars* (bold, well spaced in Steppe) *or no bars at all*; diffuse dark trailing edge seen in adult only (usually more clear-cut in adult Steppe). Some pale birds have pale primaries below with defined dark 'fingers', thus lacking pale wedge on inner primaries. Immature *often rufous- or blackish-brown on head and/or fore body, contrasting with buffish rear body*. Adult has yellow iris (dark in Steppe) and shorter gape flange than Steppe; heavy 'trousers' and heavy bill separate from spotted eagles. Fairly broad-winged with ample, deeply fingered hand (long seventh primary) and well-protruding head. Wing position in flight like Steppe. Race in Oman may be *vindhiana* (Indian Tawny Eagle). Almost identical to Tawny Eagle, but underwing-coverts very pale, often with blackish bands **Habitat** Arid mountains or plains with scattered trees; often at rubbish dumps; frequently nests on pylons. **Note** Vagrant Oman.

PLATE 23: EAGLES II

Steppe Eagle *Aquila nipalensis*   PM, WV

L: 75. W: 175–210. Adult dark brown with *uniform underwing and paler or darker flight feathers with well-spaced dark bars and clear-cut band on trailing edge* (pattern sometimes diffuse); *large dark carpal patch typical, except in darkest birds*. Above, coverts are often the palest part of the wing; usually large, dark-barred, pale primary patch (patch virtually absent in adult Eastern Imperial and Greater Spotted Eagles); grey-brown tail often boldly barred and with broader band at tip (absent in spotted eagles). *Juvenile pale brown with broad white band through underwing;* above, note large primary patch and dark rump, which separates from most young Imperial Eagles. Subadult usually has darker body than underwing-coverts, very like some immature Lesser Spotted, but told by remains of white band on underwing or well-spaced barring on flight feathers, long deeply fingered wings, ample hand (long fourth primary) *and longer, heavier bill (with peanut-shaped nostril)*. Flight heavy; often soars on flexed, flattish wings but can soar and particularly glide on arched wings with lowered hand. When perched, large heavy 'trousers' unlike the spotted eagles; *long yellow gape flange to rear of eye* separates it from other *Aquila* eagles. **Habitat** Steppe, semi-desert, hills, marshes; also rubbish dumps where hundreds may be seen. **Note** Passage and winter hatched.

Eastern Imperial Eagle *Aquila heliaca*   PM, WV

L: 72–83. W: 190–210. Adult told from Golden Eagle by *blackish-brown plumage, contrasting yellow-white hindneck, pale uppertail with broad black band and white 'braces'* (can be hard to see); also, in flight, *parallel-edged wings held flattish* and, often, *closed narrow tail when soaring*. Juvenile has *dark-streaked breast forming pectoral band which contrasts with unstreaked yellow-buff rear-body, and distinct pale wedge on inner primaries below;* yellow-brown upperparts show 1–2 complete whitish bars on coverts and *creamy lower back and rump;* lacks white band on underwing of young Steppe. Immature below, *still streaked* (much as juvenile) *or mottled blackish-brown and yellowish* with rear-body clearly paler, *possibly also retaining pale inner primaries;* adult head and tail pattern start to show early. Rather long-winged with ample hand, deep-fingered wing-tip (long seventh primary), well-protruding head and relatively long tail. Juvenile has broader, more 'S'-curved rear edge to wings. Wings sometimes slightly lifted when soaring, but arched during fast glides. Perched juveniles/immatures show *pale lower underparts* and, like adult, rather long protruding head (compared with other *Aquila* eagles). Tawny Eagle lacks streaks below of young Imperial (but seen in some African birds, which could occur in Arabia). **Habitat** Open plains and foothills with woods (nests in large or small tree); in winter also steppes, marshes, wooded desert or semi-desert, dumps. **Note** Passage and winter hatched; especially common S Oman.

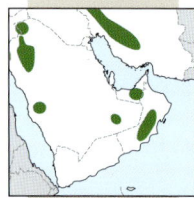

Golden Eagle *Aquila chrysaetos*   rb

L: 78. W: 190–230. Powerful flight with *flexible wingbeats; soars and glides on markedly raised wings, when note fairly long tail and slightly 'S'-curved rear edge of wings* (more pronounced in juvenile). Dark brown adult has *rusty-yellow hindneck, pale panel across upperwing-coverts,* dark-barred, black-tipped greyish flight feathers which show as *greyish area on outer wing above,* and greyish tail with blackish band at tip. (Adult Eastern Imperial Eagle is blacker above, including outer wing, has flatter, more parallel-edged wings when soaring and narrower tail.) *Juvenile and immature have white patches in primaries and inner tail, the latter with broad black band at tip,* unique in the *Aquila* eagles. Birds older than one year show pale panel across upperwing-coverts. Often hunts in tandem. **Habitat** Barren or wooded mountains, plains and semi-deserts with trees; nests in tree, sometimes on rocky ledges. **Note** Some winter dispersal; vagrant Kuwait; declining Oman.

PLATE 24: EAGLES III

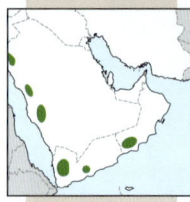

Verreaux's Eagle *Aquila verreauxii* rb
L: 80–95. W: 225–245. *Unique wing shape with very narrow base and distinct 'S'-curved trailing edge separates it from all other eagles. Adult black with conspicuous white primary patches above and below, pure white lower back and white 'V' on shoulders.* Juvenile has pale primary patches, creamy panel on upperwing-coverts, *blackish-brown throat and breast, which contrast with buffish-white rear-body* and yellowish-buff crown/hindneck. *Soars gracefully on raised wings* for long periods; when gliding on half-closed wings, narrow wing-base 'disappears'. Often hunts in tandem. **Voice** Male has loud *chorr-chorr-chorr*, female thinner *che-che-che*; also tremulous, ascending *whace-whace-whace*. **Habitat** Wild mountains; nests on cliff, rarely in tree. **Note** Formerly bred Israel; vagrant Lebanon. In Oman, rare and localised in Dhofar mountains.

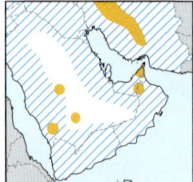

Short-toed Snake Eagle *Circaetus gallicus* PM, WV, rb
L: 64–73. W: 165–180. *Large, long-winged eagle with broad head, very pale underparts and square-cut tail with evenly spaced dark bands.* Whitish underparts variably spotted and barred; *some are nearly all whitish, others with contrasting dark head and upper breast,* lacks dark carpal patches. Flies with slow, flexible wingbeats, soars on flat or slightly lifted wings and hovers regularly. Separated from Osprey by broader wings, lack of dark carpal patch and different flight action. Pale morph Steppe Buzzard and European Honey Buzzard usually have dark carpals, blacker wing-tips, different spacing of tail-bands and are much smaller with quicker wingbeats. **Voice** Whistling, disyllabic *kee-yo* with long ascending start and short descending finish. **Habitat** Open wooded plains, stony foothills, semi-deserts; nests in a tree or on a cliff. **Note** Arabian breeders may be resident; passage hatched, a few winter in Arabia. [Alt: Short-toed Eagle]

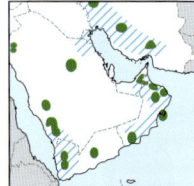

Bonelli's Eagle *Aquila fasciata* PM, WV, rb
L: 60–70. W: 150–165. In flight recalls large, thickset Honey Buzzard. *Adult identified by dark underwings contrasting with whitish underbody, pale tail with black band at tip and white patch on mantle*; at close range *white leading edge of wing*. Pale rusty-buff juvenile lacks black tail-band, and flight feathers are pale with fine dark barring; *paler translucent primaries contrast with blackish wing-tip*; when present, *narrow dark bar on rear underwing-coverts diagnostic*, but in others confined to dark 'comma' on primary coverts; upperwing cinnamon-brown with large, pale primary patch. Soars on flat or slightly arched wings, often with long, almost square-cut tail held closed (may be twisted independently); glides with carpals pressed forward, trailing edge of wings straight (recalling European Honey Buzzard). Often hunts in pairs; stoops at great speed. **Habitat** Rocky mountains, forested foothills; in winter plains and semi-deserts. **Note** Winter dispersal hatched.

Booted Eagle *Hieraaetus pennatus* PM, WV
L: 43–53. W: 110–130. Two distinct colour morphs. Size of Steppe Buzzard, but outline and wing position close to Black Kite; *tail square-cut.* More ample, deeply fingered wings than Steppe Buzzard. Pale morph has *creamy-white underparts with contrasting blackish flight feathers*, kite-like panel on upperwing, *pale scapulars (seen head-on as 'landing lights') and uppertail-coverts, and diagnostic white spots at base of neck. Lacks dark carpal patch of most pale Steppe Buzzards and European Honey Buzzards*; also has darker base to flight feathers and paler inner primaries. Dark morph similar above to pale morph but underparts dark brown, or rufous with black band through centre of underwing. When perched, *feathered tarsi also separates Booted Eagle from these and Long-legged Buzzards*. Has deeper, more powerful wingbeats and steadier glides than Steppe Buzzard; soars on flat wings; does not hover. **Habitat** Deciduous and pine forest; more open country outside breeding season. **Note** Passage hatched; a few winter in Near East and Arabia.

PLATE 25: FISH EAGLE AND HAWKS

Pallas's Fish Eagle *Haliaeetus leucoryphus*   V

L: 80. W: 190–220. A large, often vocal eagle, usually found near wetlands. Adult has *white tail with broad black terminal band*; bill dark grey (yellow in extralimital White-tailed Eagle *H. albicilla*). Juvenile told from White-tailed Eagle by *paler head with dark patch behind eye, uniform pale brown underparts, broad pale band through underwing-coverts, contrasting with dark brown leading coverts and distinct white patch or white streaks on primaries below*; up to three years old the paler head and underparts emphasise the dark eye-patch and, often, band around foreneck; underwing like juvenile but *centre of tail distinctly mottled white, forming pale band* in some; differs from young Golden Eagle by pale axillary patch and centre of underwing, paler underparts, more *parallel-edged wings held flattish when soaring* and by longer neck. When perched, bare tarsi and pale loral-patch separates Pallas's from *Aquila* eagles. **Habitat** Wetlands and rivers; also coasts. **Note** Vagrant Oman, Saudi Arabia, UAE.

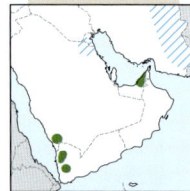

Shikra *Accipiter badius*   pm

L: 25–35. W: 50–65. Resembles rather small Eurasian Sparrowhawk in flight (blunt wing-tips) but *slightly shorter tail has rounded corners* (square-cut in Eurasian Sparrowhawk). Male of larger Iranian subspecies *cenchroides* is *pale dove-grey above, white below with faint orange barring (when close), narrow black wing-tips* and obscure pale collar. Female pale brown above with blackish subterminal tail-band, darker barring below and wing-tips barely showing black. Note *dark throat-stripe* (absent in Eurasian Sparrowhawk), grey cheeks (rufous in Eurasian Sparrowhawk) and absence of white supercilium. Smaller SW Arabian *sphenurus* male is darker blue-grey above with *contrasting black wing-tips* and lacks pale neck-collar; female shows less black on wing-tips. Juveniles (both subspecies) have *dark longitudinal spots on underparts* (largely barred in young Eurasian Sparrowhawk), banded uppertail and hardly any black on wing-tips. **Voice** Unhurried *ch-wick, ch-wick* recalls Tawny Owl; loud *kik-kooi* repeated at nest; in display a whistling *piu-piu-piu*. **Habitat** Light woodland, parks. **Note** Passage hatched; rare passage migrant and winter visitor Oman.

Eurasian Sparrowhawk *Accipiter nisus*   PM, WV

L: 29–40. W: 60–80. Female much larger than male, approaching male Northern Goshawk in size, but wingbeats faster and lighter, body slimmer, less protruding head, wing-tips blunter and tail thinner, longer and more square-cut. Adult ash-grey above (female), bluer slate-grey (male), barred rufous or brown below; whitish supercilium in female (infrequent in male); *pale underwing without dark tip*. Juvenile browner above with clear white supercilium; streaked or blotched throat and upper breast, otherwise barred below. Quick wingbeats interspersed with short descending glides (stronger, straighter glides in Northern Goshawk); display flight has slow harrier-like wingbeats, also occasionally when hunting. **Habitat** Woodland; open country with trees. **Note** Passage and winter hatched.

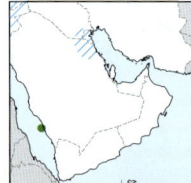

Northern Goshawk *Accipiter gentilis*   V

L: 48–60. W: 90–125. *Female much larger than male, with wingspan of Steppe Buzzard. Compared to female Eurasian Sparrowhawk male has deeper belly, slower, stronger, stiffer wingbeats and longer, broader-based, but more pointed wings*. Note shorter, broader-based tail, usually with rounded tip (thinner tail more square-cut in Sparrowhawk) and *more protruding head and neck*. Stronger, straighter glides than Sparrowhawk and often soars on upturned wings. Adult dark grey above, darker head appears 'hooded' but supercilium white; underparts finely barred. Juvenile dark brown above with pale mottling on ear-coverts; rusty-yellow *underparts boldly streaked darker*, lacks 'hooded' appearance of adult. Female told from large falcons by more rounded wings, bold tail-bands, and flight. Hunts like Eurasian Sparrowhawk but also runs down prey on ground; display flight with soft harrier-like wingbeats in shallow waves. Treated warily by crows. **Habitat** Woods, particularly coniferous, often near open country. **Note** Winter hatched, but rare Kuwait, Saudi Arabia; vagrant Oman, UAE.

PLATE 26: HARRIERS

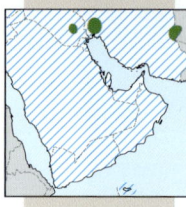

Western Marsh Harrier *Circus aeruginosus*   PM, WV

L: 48–55. W: 115–130. Larger and broader-winged than other harriers; wavering low glides on raised wings when hunting. *Male has tricoloured wings*; underwing white but wing-tip black and *rear-body red-brown*. Female dark brown with *crown, throat and breast-spot yellowish-white*. Juvenile blackish-brown, usually with rusty-yellow on head. Immature male has dirty grey areas on upperwing and tail, rusty-brown body and underwing-coverts, and more extensive black wing-tips than adult. Rare dark morph is solidly blackish-brown, but adult male has distinct white base to flight feathers below. **Voice** High-pitched Lapwing-like *vay-ee* when displaying; also *ki-ki-ki* and feeble, high 'begging' whistles. **Habitat** Marshes, reedbeds, farmland. **Note** Passage and winter hatched.

Hen Harrier *Circus cyaneus*   pm, wv

L: 45–56. W: 100–120. Slimmer than Western Marsh Harrier with more buoyant flight. Male has *clear-cut white uppertail-coverts, uniform pale grey upperparts, head and upper breast and extensive black wing-tips*. Second-autumn male can show black wedge on wing-tip like male Pallid Harrier (through primary moult). Female and juvenile brownish with white uppertail-coverts; streaked underparts whitish or rusty-yellow (warmest in juvenile); banding on secondaries below most distinct in female. Juvenile also has pale tips to greater upperwing-coverts and best separated from juvenile Montagu's and Pallid Harriers by proportionately *shorter, broader wings with more ample rounded wing-tip* (formed by four outermost primaries, but three in the other two species), *less buoyant flight and streaked breast* (unstreaked rusty yellow-brown in juveniles of the other two species). **Habitat** Marshes, meadows, farmland. **Note** Passage and winter hatched; rare E Arabia; vagrant Bahrain, UAE, Yemen.

Montagu's Harrier *Circus pygargus*   PM, WV

L: 43–47. W: 97–115. Slender build, narrow wings and buoyant flight. Male has grey back and inner wing, silvery-grey outer wing with extensive black wing-tips; one black band on secondaries above and two below; red-brown streaks below dark grey upperbreast. Second-autumn male can show black wedge at wing-tip as result of primary moult (thus recalling male Pallid Harrier). Female has rufous-streaked underparts, well-spaced dark bands across pale secondaries and evenly barred primaries from base to tip with dark trailing edge to hand; close to pale underwing-coverts and axillaries show uniform bold rufous bars. Juvenile dark rufous to yellowish-ochre below, largely unstreaked; lacks distinct pale collar of young Pallid; 'fingers' and trailing edge of hand below dark, but hand otherwise pale with fine, regular barring from base to tip. Rare melanistic morph is sooty-black with pale base to primaries below. **Habitat** Marshes, farmland; in winter/on passage any open country. **Note** Passage hatched; some winter in S Arabia.

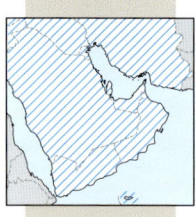

Pallid Harrier *Circus macrourus*   PM, WV

L: 40–48. W: 95–117. Proportions and flight similar to Montagu's Harrier. Male pale grey above *without clear-cut white rump; whitish head and underparts with black wedge on wing-tip*. Female from Montagu's *by pale, dark-streaked, collar* (like female Hen Harrier), *less spacing between dark bands on secondaries below, with pale bands becoming darker towards body*, primaries below often pale, contrasting with darker secondaries and lacking distinct dark trailing edge; heaviest barring is on central primaries *with bases often unbarred, creating pale 'boomerang' surrounding darkish coverts*; distal primaries with faint or no barring, except for narrow dark 'finger-tips' of longest primaries (unlike Montagu's). Except for pale leading arm, *most underwing-coverts and axillaries rather dark-streaked* and lacking distinct pattern (not bold rufous-barred as Montagu's). Streaks on underparts largely confined to upper breast, which contrasts more with paler rear-body than in Montagu's. Juvenile *has broad, pale collar bordered by brown neck*; primaries below rather evenly barred from base to tip *though often with pale 'boomerang' at primary bases, 'fingers' never all dark as in most young Montagu's*. Male (9–12 months old) has paler head and breast than Montagu's; new central tail feathers show diffuse dark bands near tips (similar Montagu's has grey neck and breast, contrasting with paler belly, and new central tail feathers plain grey). **Habitat** Steppes, grassland, agricultural fields, sandy desert. **Note** May breed Iran; passage and winter hatched.

PLATE 27: BUZZARDS

White-eyed Buzzard *Butastur teesa* V

L: 45. W: 100. Between a honey buzzard and a harrier when soaring, wings relatively narrow, held flattish, with tail half spread. When gliding wings angled and tail looks relatively long and narrow; in active flight has *Accipiter*-like wingbeats. Tail often twisted in kite-like manner. From above, adult has *cinnamon-rufous tail with black band near tip*, and warm-brown primary patch and buffish panel across wing-coverts. *Wing-tips blackish below.* At distance appears whitish below with darker breast. Young birds have paler head, whitish underparts (streaked at close range) and brown iris. Perches erect for long periods; *white throat with dark central streak and dark cheek-streak are then visible* (latter narrow or absent in young birds). Often confiding. **Voice** Plaintive, mewing *pit-weer, pit-weer*. **Habitat** Dry open country with scrub and few trees; lightly wooded foothills. **Note** Vagrant Oman.

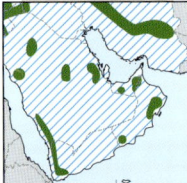

Long-legged Buzzard *Buteo rufinus* PM, WV, rb

L: 60–66. W: 130–155. Larger than Steppe Buzzard with *longer wings and tail*, kinked wing position when gliding and flexible wingbeats; *soars on raised wings*. Wide plumage variation: creamy-white, rufous-brown and blackish morphs occur, based on colour of body and underwing-coverts. The blackish morph can show coarse dark bars on flight and tail feathers. *Typical Long-legged Buzzards have pale head and breast, becoming dark towards belly, pale sandy or rufous-brown upperwing-coverts contrasting with flight feathers, unbarred pale rusty-orange uppertail and large black carpal patches*. Juvenile has finely barred outer tail and diffuse dark trailing edge to underwings; some are almost white below with bold carpal patches and dark rusty-brown belly or belly sides. Sits prominently, soars or hovers when hunting. **Voice** Recalls Steppe Buzzard. **Habitat** Steppe, semi-deserts, mountains and woodland. **Note** Passage and winter hatched. Subspecies *cirtensis* breeds Arabia; larger nominate form visits region.

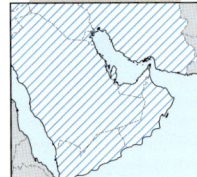

Common Buzzard *Buteo buteo* pm, wv

L: 48. W: 118. Subspecies *vulpinus* (Steppe Buzzard) occurs. Fox-red, grey-brown and rare blackish morphs occur. Smaller than Long-legged Buzzard with *shorter, narrower wings, shorter tail, stiffer wingbeats, and glides on flattish wings* (Long-legged has flexible wingbeats and kinked gliding profile). Fox-red morph has rusty-orange underwing-coverts framed by dark greater coverts, with narrow blackish comma-shaped carpal patch (patch usually large in Long-legged); rufous-brown uppertail usually more barred, head dark and pale primary patch above usually small. Juvenile has streaked breast, diffuse band on trailing edge of underwing, no broad dark band on tip of tail; often pale-headed, with pale upperwing-coverts and prominent primary patch above. Soars on raised wings; may hover when hunting. Migrates in flocks. **Voice** Mewing *peeeoo*. **Habitat** Woodlands, plains, mountain slopes with trees; anywhere on passage. **Note** Passage and winter hatched, but rarer or absent in winter.

PLATE 28: SMALL FALCONS I

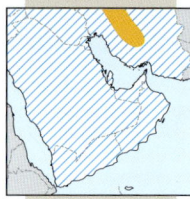

Lesser Kestrel *Falco naumanni* PM
L: 28–33. W: 63–74. Very like Common Kestrel but slightly smaller and slimmer, with slightly narrower wings, more wedge-shaped tail and quicker wingbeats. *Male is unmarked rufous above but greater coverts usually blue-grey. Head ash-grey without moustache or pale cheeks*. From below, *white underwing contrasts with dark wing-tip and creamy-buff body*, both of which have small black spots; in some, underwing-coverts virtually unmarked. Female like Common Kestrel but on average has slightly whiter, less barred flight feathers below, sometimes fewer and finer spots on underwing-coverts and greyer uppertail-coverts. *Female and juvenile can be identified by the wing-tip formula: primary 10 (outermost) longer than P8 and clearly longer than P7* or, on close views, *by pale claws* (black in Common Kestrel). Hovers less persistently than Common Kestrel, mostly taking insects in flight. Gregarious at breeding sites and on passage. **Voice** Rasping, trisyllabic *chae-chae-chae*, very different from Common Kestrel. **Habitat** Hunts over open country; nests colonially in holes in buildings, cliffs, trees. **Note** Passage hatched.

Common Kestrel *Falco tinnunculus* PM, WV, rb
L: 32–38. W: 70–78. Long, narrow, fairly pointed wings, long, slightly *tapering tail, shallow loose wingbeats, persistent hovering and rufous upperparts, contrasting with darker flight feathers*. Compared to Lesser Kestrel, *male has black spots on back and wing-coverts, lacks blue-grey greater upperwing-coverts*, and has different head pattern and more marked underwing. Female can approach male in greyness on head, tail-base and uppertail-coverts. Juvenile paler brown with thin white fringe to greater coverts. Active flight alternates with glides, some soaring and frequent hovering. *At all ages told from Lesser Kestrel by black claws and wing-tip formula: primary 10 (outermost primary) shorter than P8 and equalling P7* (useful when soaring at close range outside autumn period of primary moult). **Voice** Shrill *kee-kee-kee*, often repeated and heard mostly in breeding season. **Habitat** Open country with trees, mountains and semi-deserts; nests in hole or ledge on cliff or building; will use old nests of other species. **Note** Partial migrant.

Merlin *Falco columbarius* V
L: 25–30. W: 55–65. Female larger than male; smallest falcon in the region. *Short, pointed wings, medium-length tail*, speedy flight with fast wingbeats, interspersed with short glides. *Male told by blue-grey upperparts with blackish primaries, broad black tail-band and ill-defined head pattern*. Underparts buffish or whitish with dark streaks, or sometimes rich reddish spotting. Female and juvenile are brownish above, creamy below with dark streaks or dense dark spotting, with a *diffuse moustache, barred primaries above and five pale/dark bands of equal width on uppertail*. In steppe subspecies, *pallidus*, the male is distinctly paler blue-grey above with some rusty on neck, shoulders and mantle, and underparts are whiter. Female and juvenile *pallidus* are rufous above with Common Kestrel-like dark bars (but kestrel's flight, proportions and denser tail-barring prevent confusion). Hunts usually low over ground with undulating flight, changing direction, followed by a straight attack. When perched, wings fall well short of tail tip. **Habitat** Open country; steppes and semi-deserts, marshes, farmland and plains. **Note** Passage and winter hatched; rare SW Arabia; vagrant Bahrain, Oman.

PLATE 29: SMALL FALCONS II

Amur Falcon *Falco amurensis*   pm, wv
L: 26–32. W: 65–75. A small falcon, smaller than Eurasian Hobby which it resembles in flight silhouette. Adult male is uniformly *dark grey with conspicuous, pure white underwing-coverts, red thighs and undertail-coverts, grey tail and paler grey underparts than upperparts. Bill and legs bright orange.* (Closely resembles **Red-footed Falcon** *F. vespertinus* – not yet recorded in Oman – and separated from that species by white underwing-coverts.) Adult female is white below with warm *buff thighs and streaked breast, boldly barred flanks* and *lightly spotted underwing-coverts; dark crown and short moustache contrast with white cheeks.* Juvenile and immature resemble Eurasian Hobby but note white ground-colour below and darker crown. White underwing-coverts appear on male at one year of age. Female and juvenile told from Eurasian Hobby by barred undertail, smaller moustache, paler underside, flight and hovering. **Habitat** Cultivation, lightly wooded areas. **Note** Passage hatched, uncommon, but regular in S Oman; vagrant Kuwait, Qatar, Yemen.

Eleonora's Falcon *Falco eleonorae*   V
L: 39. W: 97. *Long-winged, long-tailed,* recalling Eurasian Hobby but larger. Flight swift and agile or *relaxed with slow wingbeats*. Pale morph recalls Eurasian Hobby but has *darker underparts and dark, unmarked underwing-coverts contrasting with pale-based, unbarred flight feathers*; at distance looks dark below except for pale throat and cheeks. Dark morph (25% of population) *uniform blackish-brown*; from male Red-footed Falcon by size, proportions, flight, underwing-pattern, dark primaries above and lack of red thighs. Juvenile (both morphs) paler below than adult pale morph. *Told from Eurasian Hobby by dark underwing-coverts contrasting with paler flight feathers, which have dark trailing edge*; also thinner moustache. Often hunts in flocks, especially at dusk; catches insects in flight, sometimes hovers or stoops. Breeds as Sooty Falcon. **Voice** Loud hoarse *kjie-kjie-kjie* when breeding. **Habitat** Rocky islands and sea cliffs; often hunts over wetlands. **Note** Vagrant Oman, E Saudi Arabia, UAE.

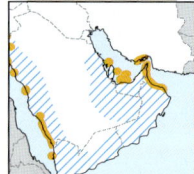

Sooty Falcon *Falco concolor*   MB
L: 32–38. W: 85. Larger than Eurasian Hobby and long-winged like Eleonora's Falcon, but tail slightly shorter, with elongated central tail feathers. Adult from dark morph Eleonora's by *slaty-grey upperparts with darker primaries and outer uppertail; blue-grey underparts* (underwings paler) *without Eleonora's contrasting underwing pattern*. Female darker than male. Juvenile from similar Eurasian Hobby and pale morph Eleonora's (adult and juvenile) by *greyer upperparts with darker wing-tip, less clearly streaked underparts, the spot-streaks almost merging on upper breast; lightly marked dusky underwing* has dark wing-tip and trailing edge (Eleonora's has dark coverts contrasting with paler flight feathers; Eurasian Hobby has uniform underwing); *undertail finely barred except near tip* (Eurasian Hobby and young Eleonora's have undertail barred to tip). Flight recalls Eurasian Hobby but glides on level wings. When perched, wing-tip reaches tail-tip or slightly beyond. Breeds late summer, feeding young on autumn migrants. **Voice** Shrill ringing alarm *kee-kee-kee* at nest site. **Habitat** Colonial on islands, inland desert cliffs; usually nests in hole. **Note** Passage hatched; vagrant Kuwait.

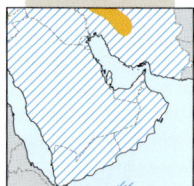

Eurasian Hobby *Falco subbuteo*   PM
L: 32–36. W: 74–92. *Scythe-like, pointed wings* and relatively short tail. Adult has *slate-grey upperparts, uniform tail, prominent moustache and conspicuous white cheeks*, densely streaked underparts with *red thighs and undertail-coverts*. Juvenile browner above with pale feather fringes, lacks red thighs and undertail-coverts; told from juvenile Red-footed Falcon by more distinct breast streaking, darker head with more contrasting face-mask, unbarred uppertail and underwing pattern. Flight swift and agile; has strong steady wingbeats, short fast glides; accelerates when hunting birds, soars when catching insects; rarely hovers (and only briefly). **Habitat** Scattered woodland, cultivation, open country. **Note** Passage hatched.

PLATE 30: LARGE FALCONS

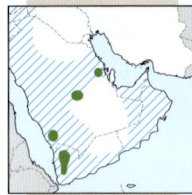

Lanner Falcon *Falco biarmicus* — pm, wv

L: 42–52. W: 95–115. Peregrine-sized, resembling Saker Falcon in plumage and shape; long wings, slightly blunt-ended when soaring; tail relatively long. *Adult from Saker by barred, greyish upperparts, distinctly barred uppertail*, more contrasting head pattern: *black forehead band, clear-cut narrow black eye-stripe, conspicuous moustache and spot-bars on flanks. Crown unstreaked creamy-buff* (Middle Eastern *tanypterus*), *chestnut* (NE African *abyssinicus*) or *pale rufous rear crown and nape* (European *feldeggii*). Juvenile dark brown above with boldly streaked underparts and rear underwing-coverts; *unbarred closed uppertail* (unlike most Sakers). Contrasting underwing pattern and more densely streaked underparts separates it from juvenile Peregrine and Barbary Falcons. Moderately slow, stiff wingbeats, faster when hunting; stoops or runs down prey; soars with wings level or slightly upcurved. **Voice** Slow, scolding *kraee-kraee-kraee* at breeding site. **Habitat** Mountains, plains and semi-deserts. **Note** Declining; dispersal hatched, but rare; escapes confuse true picture.

Saker Falcon *Falco cherrug* — pm, wv

L: 47–55. W: 105–125. Like Lanner Falcon but larger, heavier-chested with *creamy crown* (sometimes just nape), *unbarred kestrel-like contrast above, less distinctly barred uppertail, poorly developed moustache and less contrasting head pattern* (diffuse eye-stripe, no dark band on forehead). Whitish supercilium often more conspicuous and belly more spotted, but lacks Lanner's spot-bars on flanks. Sakers are greyish above with dark bars, including uppertail; these are *best told by head pattern and size*. Juvenile similar to young Lanner, but outer tail feathers generally conspicuously spotted buff on outer webs (seen well in half-spread tail) and dark stripe behind eye less clear-cut; best told by size. When perched, *wing-tip falls short of tail-tip* (unlike most Lanners) and 'trousers' heavier, covering much of tarsus. Slow, flattish wingbeats; when soaring, wings flat or slightly upcurved. **Voice** Like Peregrine, but harsher; also thin, querulous note like cross between Curlew and Herring Gull of W. Europe. **Habitat** Wooded steppes, foothills, mountains, semi-deserts. **Note** Passage and winter hatched; declining and rare in all areas.

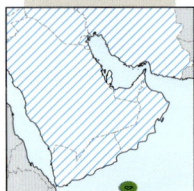

Peregrine Falcon *Falco peregrinus* — PM, WV

L: 40–52. W: 85–120. Large, stocky with relatively short tail and *broad-based, sharply tapering wings, more pointed than Saker and Lanner* when soaring. Adult told by *black crown and bold moustache, contrasting with white throat and cheeks, barred underparts, white upper breast and uniform underwing*. Juvenile *has smaller whitish cheek-patch* (not reaching eye, unlike Saker and Lanner) and also uniform underwing (unlike Saker and Lanner). Young of migrant *calidus* are tricky, showing Saker-like head pattern and large size, but told by underwing pattern and wing shape. Subspecies breeding in region, *brookei*, is more compact, like Barbary Falcon, with salmon wash to breast and sometimes rufous wash on nape. Fairly quick, shallow, stiff wingbeats; impressive when hunting, with long, fast stoops. **Voice** Alarm loud, scolding *aack-ack-ack*. **Habitat** Mountains, forests, cliffs; outside breeding season also marshes, wastelands. **Note** Passage and winter hatched.

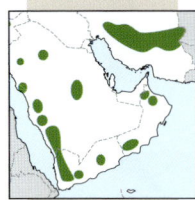

Barbary Falcon *Falco pelegrinoides* — PM, rb

L: 32–45. W: 80–100. Resembles Peregrine, especially *brookei*, but slightly narrower-based wings give impression of longer tail. Adult told by *rufous nape and rear eyebrow, narrower moustache and larger pale cheek-patch, almost reaching eye; more creamy, less barred underparts*, confined to flanks in E. Iranian *babylonicus*, which has redder crown; *underwing whiter with more extensive dark wing-tips* than Peregrine; *often with dark 'comma' on greater primary coverts*; upperparts paler blue-grey, with darker end to tail. Juvenile like Peregrine, *but narrower moustache, larger cheek-patch, tawny supercilium and rusty nape*; rustier underparts with thinner, more restricted streaks. Some juveniles have yellow cere and legs soon after fledging (in Peregrine blue-grey, usually becoming yellow in first winter). From young Lanner by pattern of underwing and underparts. **Voice** Harsh *keck-keck-keck*, less hacking than Peregrine. **Habitat** Arid mountains, semi-deserts. **Note** Some autumn and winter dispersal; vagrant Kuwait, Qatar.

PLATE 31: BUSTARDS AND THICK-KNEES

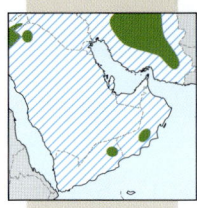

Macqueen's Bustard *Chlamydotis macqueenii* — rb, pm, wv
L: 60. W: 150. A large bustard, with *black frill down side of neck in all plumages*. Flight rather slow, wingbeats shallow, with *long-tailed and narrow-winged appearance* and *white patch confined to outer primaries only*. Shy, prefers sneaking away without flying. **Habitat** Stony or sandy steppes, semi-desert; also marginal cereals and other crops. **Note** Formerly more widespread breeding range; has bred Kuwait, Syria; may breed Iraq; reintroduced in some sites in Saudi Arabia and UAE. Now separated from Houbara Bustard *C. undulata*, which occurs across North Africa.

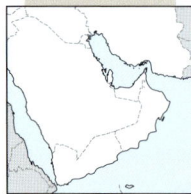

Little Bustard *Tetrax tetrax* — V
L: 43. W: 110. Small, with *small head and long neck*. Male has *grey, black-and-white neck pattern*; non-breeding male, female and juvenile lack these striking neck markings. Flight fast with rapid, stiff, winnowing wingbeats showing *almost completely white wings with black mainly confined to outermost primaries*. Male displays in spring with inflated neck and brief leaps in the air. Often in flocks, especially in winter; rather shy, frequently in cover of grass or low vegetation. **Voice** Male displays with a short *prrrt* call. In flight, male's wings make whistling noise. **Habitat** Grassy plains, large cereal fields or fodder crops. **Note** Formerly bred Syria; winter hatched, but rare Turkey; vagrant Oman.

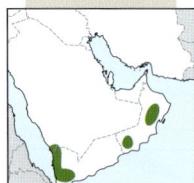

Spotted Thick-knee *Burhinus capensis* — RB
L: 43. Resembles Eurasian Stone-curlew but upperparts buffier, *spotted black in adult* (but streaked in juvenile like Eurasian Stone-curlew), the spotting being particularly obvious on the paler coverts; *lacks black-and-white bars on coverts. In all ages tertials and tail are diagnostically barred*. Flight pattern shows two prominent white patches on black primaries, and underwing usually shows a strong dark bar along central wing. Prefers to stay near cover of bushes. More active by night than day. **Voice** Usually at night: a whistled *ti-ti-ti-tee-tee-tee ti ti ti*, growing to a crescendo, then dying away. **Habitat** Savanna and scrub, rocky river beds, broken ground, more bushy habitat than Eurasian Stone-curlew frequents; nests near cover of bushes. **Note** Breeds C and S Oman; rare elsewhere in Oman. [Alt: Spotted Dikkop]

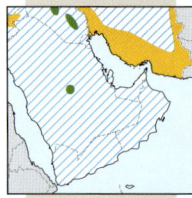

Eurasian Stone-curlew *Burhinus oedicnemus* — pm, wv
L: 42. W: 81. Large, streaked, curlew-coloured wader with *short bicoloured bill and large, staring yellow eyes*. Found in dry habitats, often 'frozen' motionless or walking slowly in hunched posture. Flies with *stiff wingbeats and shows two small white 'windows' in primaries and a paler wing-panel, bordered in front with a dark and a white line*. Adult Spotted Thick-knee is darker and boldly spotted (juvenile more streaked). Often encountered in flocks. More active by night than day. **Voice** Vocal mainly at night, reminiscent of Eurasian Curlew's *cur-lee* with emphasis on higher-pitched second syllable; also loud Oystercatcher-like *ku-beek, ku-beek*. **Habitat** Open plains, steppe and semi-desert, also extensive arable land; among scattered trees and light scrub in hotter climates. **Note** Passage and winter hatched, but absent in winter in Iran and Turkey.

Great Stone-curlew *Esacus recurvirostris* — pm, wv
L: 50. W: 95. Larger than Eurasian Stone-curlew with plain sandy-grey upperparts, white underparts with *unstreaked grey neck and breast*. Head distinctive: *long, very heavy, upturned, yellow-based black bill and striking black-and-white head markings*. In strong flight, short-tailed appearance recalls small goose; flight feathers black with striking white patches in primaries; wing-panel greyish, contrasting with dark band on lesser coverts, also visible on closed wing. Runs fast. **Voice** Territorial note a wailing whistle with a rising inflection, mostly at night, *see* or *seeey*; alarm call a harsh *see-eck*. **Habitat** Rocky river beds and their barren environs, coastal reefs, beaches, estuaries and saltpans. **Note** Winter hatched, but rare. [Alt: Great Stone Plover]

PLATE 32: WATER RAIL AND CRAKES

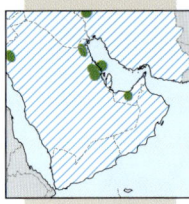

Water Rail *Rallus aquaticus*   pm, wv
L: 26. W: 41. Secretive; noticeably smaller than Common Moorhen, but *larger than the* Porzana *crakes* and easily told by *slender, long red bill*. Adult has dark, *mottled brown upperparts*, uniform slate-blue sides of head, and underparts wit*h heavy black-and-white bars on flanks and conspicuous white undertail*. Juvenile has browner face, mottled grey-blue underparts and blackish bill. Tail often cocked and jerked when walking. When flushed, flies only a short distance on fluttering, rounded wings with long legs dangling. **Voice** Grunting, groaning, whining and *stomach-churning sounds* from vegetation; sometimes *like a squealing pig*. In spring, male (and female) utters for hours a rhythmic *trüt-trüt, trüt*, sometimes ending with a trill. **Habitat** Dense aquatic vegetation, ponds, ditches. **Note** Passage and winter hatched.

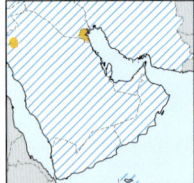

Little Crake *Porzana parva*   pm, wv
L: 19. W: 37. Smaller than Spotted Crake and separated by *uniform blue-grey underparts (adult male) and heavily barred undertail-coverts*. Less compact than Baillon's Crake with longer legs and neck, and *much longer wing-projection*. Tertial pattern differs: pale buff fringes to inner webs form *broad creamy line along inner aspect of folded wing* (in Baillon's paler edges to tertials never form broad continuous line). Male also told from Baillon's by *less barring on flanks*, less spotted upperparts, red base to bill, and green legs. *Female has brown-buff underparts*, white chin and throat and some grey on cheeks and supercilium. Juvenile lacks grey head pattern and has stronger flank-barring than adult; told from juvenile Baillon's by *structure, tertial pattern* and less barred underparts. (Any small crake with buff underparts seen in the region in mid-winter and spring will be female Little Crake.) **Voice** Male's song loud, accelerating croaking *kwak... kwak... kwak, kwak, kwak-kwak-kva-kva-kva-kva-kva*. Female's call short, accelerating, with vibrant terminal trill *kwek, kwek-kverrrrr*. **Habitat** Swamps and wetland fringes; fondness for high reeds in deeper water and lagoons with floating vegetation. **Note** Passage hatched, but rare throughout much of region; some winter Iraq and S Arabia.

Baillon's Crake *Porzana pusilla*   PM, WV
L: 18. W: 35. Resembles Little Crake but *more compact and with very short primary projection*. Upperparts warmer rufous with distinct, but small, irregularly scattered white spots; *tertials do not form continuous pale line along inner aspect of folded wing* as in Little Crake; underparts bluish-grey in both sexes with *heavier black-and-white barring on flanks*. Uniform green bill without *red base*, and dirty olive legs and feet. Juvenile more strongly barred below than Little Crake and best told by short primary projection and absence of broad, pale tertial-line. Most skulking of all the crakes. **Voice** Song a series of dry, rattling frog-like sounds lasting 1–2 seconds and repeated at intervals of 1–2 seconds, *trrrrr, trrrrr, trrrrr*. **Habitat** Dense vegetation (sedges, rushes), small pools, wetland edges. **Note** Passage hatched, but rare or sometimes vagrant; some winter in S Arabia.

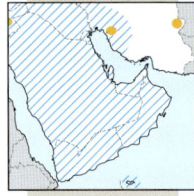

Spotted Crake *Porzana fusca*   PM, WV
L: 23. W: 40. Small, round-bodied crake, slightly larger than Little Crake. Note *heavily white-spotted plumage and* short, red-based, yellow bill; flanks and vent strongly barred black and white, but *buff undertail-coverts* (barred in Little and Baillon's Crakes) visible when walking with tail cocked. Juvenile lacks grey head pattern, has whitish throat and bright brownish underparts with whitish spots. Secretive, moves with slow, stalking steps and sudden crouching run. In short flight often dangles legs. **Voice** Song (both sexes) is a *rhythmical, far-carrying whistle* whitt, whitt, repeated each second, mainly from late dusk and through the night. **Habitat** Swamps, overgrown ditches, margins of ponds. **Note** Passage hatched, some winter.

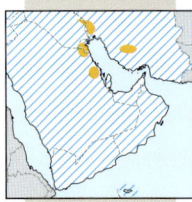

Ruddy-breasted Crake *Porzana fusca*   V
L: 22. W: 37. Adult has *rich chestnut underparts, dark brown upperparts*. Black-and-white barring on lower belly and undertail-coverts. Legs and eyes red, bill dark grey. Sexes alike but female paler with whiter throat. Juvenile has darker upperparts and duller underparts; chin and throat white becoming more chestnut with age. **Voice** Soft *puk* while foraging. **Habitat** Wetlands, edges of water, dense vegetation. **Note** Vagrant Oman.

PLATE 33: CORN CRAKE, WATERHENS AND SWAMPHENS

Corn Crake *Crex crex*   pm
L: 29. W: 50. Skulking, rarely seen out of cover unless flushed; larger than Common Quail. In flight can suggest young gamebird, but note *chestnut wing-coverts* and often dangling legs. Otherwise located by *characteristic song* in breeding season. **Voice** Male's breeding call a far-carrying, disyllabic rasping, *arrp-arrp, arrp-arrp*. **Habitat** Meadows, lush vegetation and crops; avoids standing water. **Note** May breed Iran; passage hatched, but rare; some winter in S Arabia.

White-breasted Waterhen *Amaurornis phoenicurus*   pm, wv
L: 32. W: 49. Slimmer than Common Moorhen; *slaty-brown and white, with rufous-chestnut vent and undertail-coverts*; crimson eye set in white face is very conspicuous; in breeding season note reddish base to upper mandible. Juvenile has white face obscured by slate-brown. Skulking, but often easily seen in open, walking with jerking tail, displaying undertail-coverts. Climbs in tangles and creepers, sometimes into canopy. Flies with dangling legs; occasionally swims. **Voice** Very vocal in breeding season; call a prolonged ululating *kaargh-kaargh*, or breaking into *kurrwah-kurrwagh-kurrwagh*, *krrr-kwok-kwok-krr-oowark-oowark*. **Habitat** Ponds and tangled cover. **Note** Winter hatched, but rare; vagrant Qatar, Saudi Arabia, Yemen.

Grey-headed Swamphen *Porphyrio poliocephalus*   V
L: 47. W: 95. Large and heavy; twice the size of a Common Moorhen. *Uniformly bluish-purple with grey head and neck, huge red bill and frontal shield; red legs and long toes.* Swims with body tilted forward, *white undertail-coverts often striking*. In flight like huge Common Moorhen but with blue wings and long red legs. Juvenile drabber, with greyish-blue underparts, dull red legs and greyish bill. Shy and often in cover. **Voice** Loud clucks, clanks, bugling and deep mooing notes; a low *chock-chock*; also *tschak-tschak* and nasal bugled song *quin quin krrkrr, quin quin krrkrr*. **Habitat** Swamps with extensive reedbeds, borders of lakes fringed with tall, dense cover. **Note** Vagrant Bahrain, Oman. [Formerly Purple Swamphen *Porphyrio porphyrio*, from which now separated]

African Swamphen *Porphyrio madagascariensis*   V
L: 47. Similar to Grey-headed Swamphen, but body *and head bluish-purple, with mantle and much of wings (scapulars and tertials) clearly dirty moss-green*. Immatures duller with lighter greyish-blue underparts, though still greenish (often muted) on upperparts. **Voice** Similar to Grey-headed Swamphen. **Habitat** As Grey-headed Swamphen. **Note** Vagrant Oman; escapes apparently bred recently in UAE. [Formerly Purple Swamphen *Porphyrio porphyrio*, from which now separated]

Watercock *Gallicrex cinerea*   V
L: Male 43, female 36. Larger than Common Moorhen. Male in breeding plumage blackish with pale feather-edgings and *bright red frontal shield and legs*. Female, non-breeding male and immature *buff-brown streaked darker above* and narrowly banded below; frontal shield and bill yellowish; legs brownish-green. Mainly seen at dawn and dusk, never far from cover. Jerks tail; in flight, dangles legs. **Voice** Generally silent outside breeding season. **Habitat** Swamps, marshes and, on migration, along coasts. **Note** Vagrant Oman.

PLATE 34: GALLINULES, MOORHENS AND COOTS

Allen's Gallinule *Porphyrio alleni*   V
L: 23. W: 50. Smaller, more elegant than Common Moorhen, with bluish head and underparts, and *iridescent greenish wings. Lacks white flank-line,* but has white undertail-coverts. Note *red legs and greenish upperparts*. Juvenile/first-winter has warm brown upperparts, distinctly pale-fringed tertials and greenish wash to flight feathers; underparts buffish with white belly and undertail-coverts; frontal shield brownish; legs brownish turning red. Jerks tail when moving, and walks easily over floating plants; swims well. Long legs dangle in short flight. Secretive. **Habitat** Marshes, nearby rank grass or thick bushes. **Note** Vagrant Oman.

Common Moorhen *Gallinula chloropus*   RB, WV
L: 33. W: 52. Dark, hen-like waterbird, with prominent *red bill and shield, and white flank-line. Constantly jerks tail to show white undertail-coverts.* Swims with vigorous nodding movements and body tilted forward. Juvenile paler, grey-brown with dark bill; distinguished from young Eurasian Coot by *white flank-line and undertail pattern*. Seeks cover readily and patters over water when disturbed; often walks openly in wet grassland, marshes. **Voice** Sings at night with persistent clucking *kreck-kreck-kreck*. Many calls can be confused with those of Eurasian Coot; sometimes a short variable kek or *kr-r-eck*; also characteristic sudden, loud, gurgling, *grrll* and a 2- or 3-note *kwett, kwette-wett*. **Habitat** Freshwater wetlands and edges of pools with cover. **Note** Passage and winter hatched.

Lesser Moorhen *Gallinula angulata*   V
L: 28. Smaller than Common Moorhen and told from it in adult plumage by yellow bill and absence of red at top of legs. Juvenile very similar to juvenile Common Moorhen but differs in having dull yellow bill (dark in Common Moorhen) and *paler head and underparts,* the latter contrasting with the brown upperparts (more uniform plumage in juvenile Common Moorhen). Habits similar to Common Moorhen. **Habitat** As Common Moorhen but also areas of temporary water. **Note** Vagrant Oman.

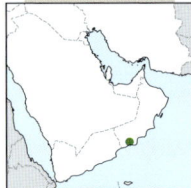

Red-knobbed Coot *Fulica cristata*   rb
L: 39–44. Adult very similar to Eurasian Coot, but differs in less rounded head shape, variably dusky grey-blue bill but white frontal shield and *two prominent red knobs on forehead* in breeding season; these are shrunken and inconspicuous at other times. Immature and winter adult best identified by shape of *feathering at bill base, being rounded in Red-knobbed* but a sharply pointed wedge in Eurasian Coot. In flight wings lack white trailing edge seen in Eurasian Coot. **Warning:** Hybridisation with Eurasian Coot can occur, the offspring showing characters of both species. **Voice** Shrill nasal *kerre, krrk* or rolled *krre-krre-krre*, quite unlike calls of Eurasian Coot. **Habitat** Marshes, lakes and lagoons with cover close by. **Note** Vagrant Oman, UAE. [Alt: Crested Coot]

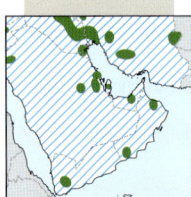

Eurasian Coot *Fulica atra*   WV, pm, rb
L: 36–42. W: 75. *Adult sooty-black with white bill and frontal shield*; hunch-backed on water. Upright stance out of water when note *long greenish legs and lobed feet*. Flight stronger and heavier than other rails, more duck-like, on rounded wings and with long pattering run across water before take-off, with *narrow white trailing edge on inner wing*. Long toes trail behind tail-tip. Juvenile duller and paler with nearly white underparts and smaller frontal shield. Dives well, but only for a short time. Markedly gregarious, especially in winter. **Voice** Commonest call is a short staccato *kewk*, also an explosive high *pitts*. **Habitat** Lakes, reservoirs, ponds with grassy margins; sometimes saltwater in winter. **Note** Passage and winter hatched.

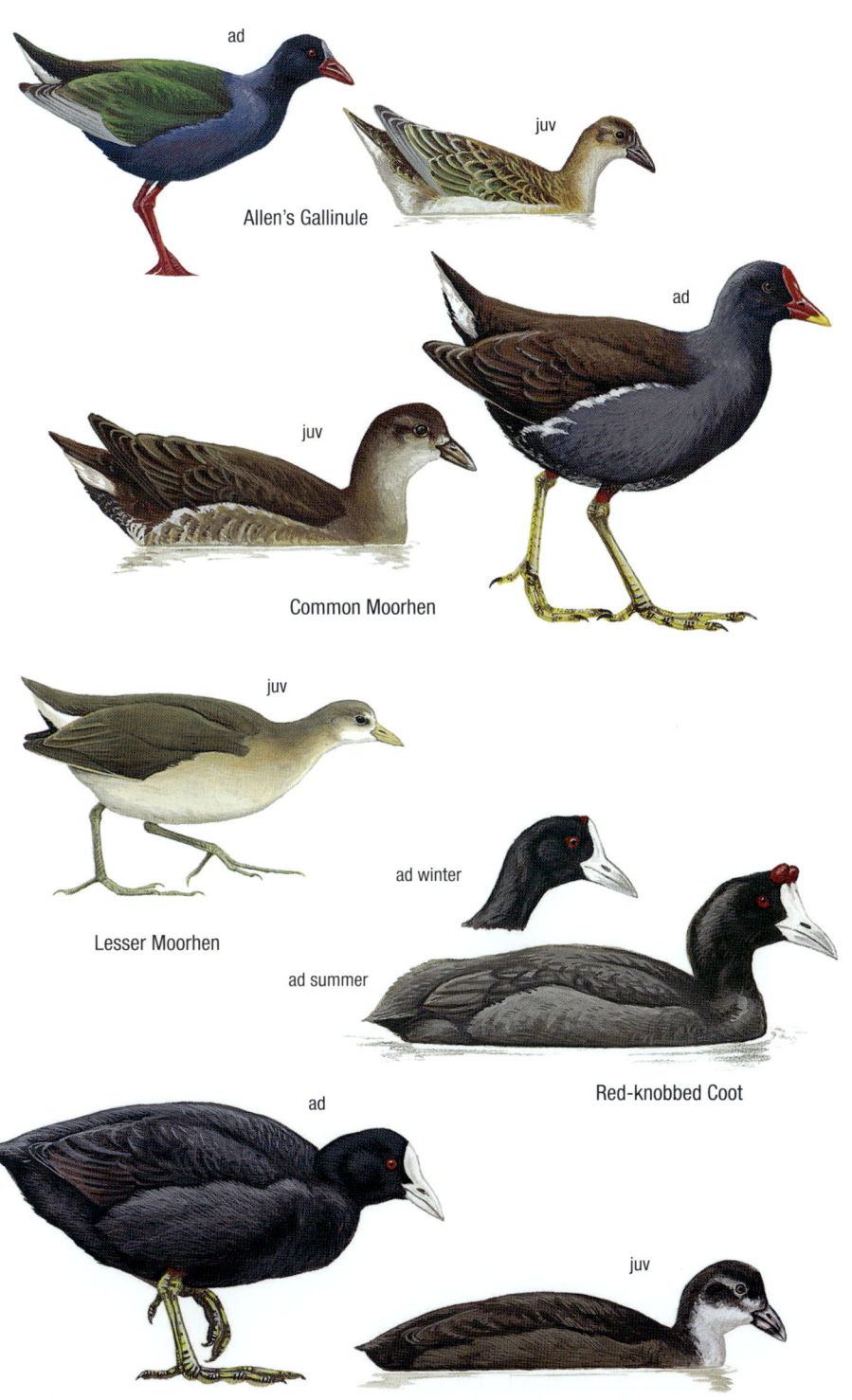

PLATE 35: LARGE PIED SHOREBIRDS

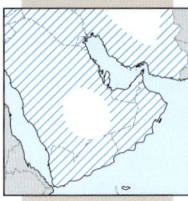

Eurasian Oystercatcher *Haematopus ostralegus* — PM, WV
L: 43. W: 83. Large *black-and-white wader with long, red bill and rather short, red-pink legs (adult)*. Flight strong and direct; shows *conspicuous white wing-bar*, white rump and *terminal black tail-band*. Non-breeding adult has white neck collar and duller bill-tip. Juvenile and immature have duller black upperparts, dark tip to bill, greyish-pink legs and white neck collar. Gregarious. **Voice** Noisy; common call far-carrying *kleep-kleep* and a disyllabic *pick-pick*. **Habitat** Open areas near lakes and rivers; in winter mainly coastal. **Note** Passage hatched, rare inland; winters on all coasts, some remaining in summer.

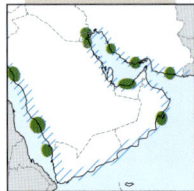

Crab-plover *Dromas ardeola* — WV, rb
L: 39. W: 77. Distinctive with *large head and straight, massive bill*. Adult winter and immature have dark streaks on crown and nape. Juvenile shows streaking on rear crown, greyish mantle, scapulars and tail, and at distance could be mistaken for a gull. *In flight, shows white wings with blackish flight feathers and long trailing legs*; flight slow with stiff wingbeats and head sunk into shoulders. Often gregarious. **Voice** Noisy; shrill *tchuk-tchuk-tchuk* near nest; alarm a sharp *kjep*; in flight *chee-rruk*, often by night. **Habitat** Coasts, mudflats, coral reefs; never inland. Nests in tunnel excavated in sandy ground; colonial. **Note** Disperses to hatched areas.

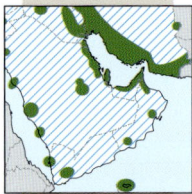

Black-winged Stilt *Himantopus himantopus* — RB, PM, WV
L: 37. W: 75. *Exceptionally long, deep-pink legs and slim black-and-white body*; walks with high, graceful carriage. *In flight, white with uniform black wings and obvious long trailing legs*. Female has slightly browner mantle and scapulars. Non-breeding adults have dusky head and neck marks. Juvenile and immature brownish above with greyish crown and hindneck and white trailing edge to wings in flight. **Voice** Noisy in breeding area; a variable sharp *kek*, high-pitched continuous *kikikikik* or *kee-ack*. **Habitat** Shallow fresh or brackish water, estuaries. **Note** Range expanding; passage and winter hatched.

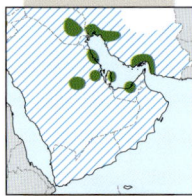

Pied Avocet *Recurvirostra avosetta* — wv, pm
L: 44. W: 79. Slender, pied wader with mainly white appearance. *Black crown and hindneck and long, thin, upcurved bill* distinctive. Flight stiff-winged and flickering, showing *white wings with prominent black markings* and long trailing grey-blue legs. Juvenile tinged brown on black parts, and white is mottled buff on upperparts. Walks steadily and delicately with head down when feeding in shallow water. Gregarious. **Voice** Noisy on breeding grounds; most common call a repeated *bluit-bluit*. **Habitat** Saline, mudflats, estuaries and sandbanks; nests colonially near shallow water. **Note** Passage and winter hatched.

PLATE 36: LAPWINGS

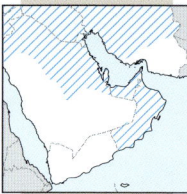

Northern Lapwing *Vanellus vanellus* pm, wv
L: 30. Unmistakable with *long, thin, upturned crest and greenish upperparts*. Underparts clean white with broad black breast-band and rich burnt-orange undertail. *In rather flappy flight, broad, rounded wings show no wing-bar*. In winter, upperparts have narrow pale scaling to feathers. Juvenile lacks tall crest, instead having scruffy, spiky tuft on nape. Often in large flocks outside breeding season. **Voice** Loud, shrill *peeo-vit* uttered in tumbling display flight over breeding grounds. **Habitat** Open fields, marshes, shallow pools and coastal flats. **Note** Passage and winter hatched, scarce in south.

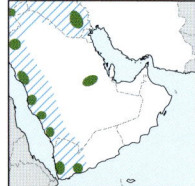

Spur-winged Lapwing *Vanellus spinosus* pm, wv
L: 26. W: 75. Elegant, long-legged plover with *black head and underparts and contrasting white cheeks and sides of neck*. In flight, tricoloured wing pattern: conspicuous white band between black flight feathers and sandy wing-coverts; also broad black tail-band; from below black belly and flight feathers contrast with white underwing-coverts. Juvenile similar to adult. Legs black. **Voice** Noisy in breeding area; alarm call is a shrill Oystercatcher-like *dwitt-dwitt* or *kwitt-kwitt*. **Habitat** Fresh and saline marshes, irrigated land, with short vegetation. **Note** Dispersal and passage hatched; vagrant Gulf States.

Grey-headed Lapwing *Vanellus cinereus* v
L: 35. W: 75. Grey head and neck, black lower breast separating neck from white belly. Brown back. Bill distinctive: yellow with black tip. Eyes red, legs yellow. In flight has upperwing pattern similar to White-tailed Lapwing, but tail white with black patch like Sociable Lapwing. **Voice** Flight call a sharp *kick-kick*. **Habitat** Wet grasslands and fields, marshes. **Note** Vagrant Oman, from E Asia.

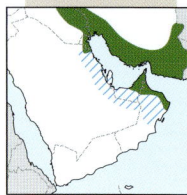

Red-wattled Lapwing *Vanellus indicus* RB
L: 33. W: 80. Rather large, colourful plover; easily identified by *red bill, eye-ring and wattles, bright yellow legs, black head and centre of breast*. Flight light with slow wingbeats, showing similar wing pattern to Spur-winged and White-tailed Lapwings, but *tail has black subterminal band with broad white terminal band*; yellow feet project distinctly beyond tail. Juvenile much duller with chin and throat almost white, black areas grey-brown, and wattle is tiny or absent. Fairly confiding. **Voice** Noisy; loud and shrill alarm notes rendered as *did-he-do-it, pity-to-do-it*. **Habitat** Open country, usually near fresh water, and showing preference for grassy fields and agricultural land. **Note** Dispersal as hatched.

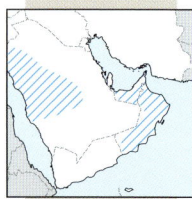

Sociable Lapwing *Vanellus gregarius* wv, pm
L: 29. Lapwing-sized, with more upright stance, especially when alert. In breeding plumage easily told by *long white supercilium, joining on nape, black crown and chestnut-black belly*. In winter, loses belly-patch, becomes mottled on breast and supercilium is less distinct (but white forehead usually quite prominent). Juvenile, which is browner than adult, has buffish wash to forehead and supercilium, and pale feather-edgings on upperparts. Legs dark. In flight, shows fairly rounded wings, with conspicuous black, white and brownish upperwing pattern and black band on tip of tail. Typical plover actions on the ground, making short runs with head tucked into body, stopping to peck at ground or stand with head erect. Often in flocks, on passage and in winter sometimes with Northern Lapwings, golden plovers and coursers. **Voice** Harsh *chark-chark-chark* flight call. **Habitat** Steppes and bare or cultivated fields (e.g. with winter cereals); rare on coast. **Note** Passage and winter hatched; vagrant Bahrain, Kuwait, Yemen.

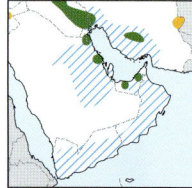

White-tailed Lapwing *Vanellus leucurus* WV, pm
L: 28. Slender and graceful, most closely resembling Sociable Lapwing in winter plumage but smaller and readily told by *plain head* (accentuating dark eye), *longer, deep waxy-yellow legs* (which protrude in flight) and, *in flight, all-white tail*. Juvenile paler on neck and breast, dark-mottled on upperparts with dark cap and faint brown tip to tail. When feeding, tips down so steeply that it almost stands on its head! **Voice** High-pitched *kee-vee-ik*, persistently repeated at breeding sites. **Habitat** Fresh or saline pools, marshes and wet plains; may nest colonially. **Note** Range expanding; passage and winter hatched.

PLATE 37: LARGE PLOVERS AND JACANA

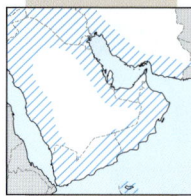

Grey Plover *Pluvialis squatarola*   PM, WV
L: 29. W: 77. Larger than the golden plovers with *heavier head and longer, stouter bill*. *Greyish appearance* lacking obvious yellow or greenish tones on upperparts, except for faint yellow-buff tinge in juvenile. *Black axillaries diagnostic in flight (all ages)*. Breeding plumage recalls Eurasian Golden, but more white on head and nape and *coarsely speckled black-and-white upperparts*; larger white breast-side patch *without white flank-line*. *White wing-bar and rump* obvious in flight. **Voice** Flight call a mournful trisyllabic whistle, *dee-oo-wee* (second note lower-pitched), often repeated. **Habitat** Tidal flats and saltings; occasionally inland wetlands. **Note** Passage hatched, rare inland, winters all coasts, with some remaining in summer. [Alt: Black-bellied Plover]

Pacific Golden Plover *Pluvialis fulva*   PM, WV
L: 24. W: 66. Similar to slightly larger Eurasian Golden in all plumages. *Best separated by slimmer build, dark underwing and voice*. Compared to European Golden, wings often protrude 1–2cm beyond tail-tip, bill is finer and longer, legs are clearly longer, particularly thighs, making it appear more elegant. *In flight, toes extend beyond tip of tail and wings appear longer and narrower; underwing and axillaries are greyish-brown in all plumages* (white in European Golden). Adult in breeding plumage brighter than European Golden, more golden on mantle and scapulars and often bright white spangling on the wing-coverts; non-breeding and immature have more distinct yellow-buff supercilium than European Golden. Upperwing as European Golden. **Voice** Soft disyllabic call *gru-it* (resembling Spotted Redshank). **Habitat** Mudflats, grassland, cultivated fields. **Note** Passage and winter hatched.

European Golden Plover *Pluvialis apricaria*   pm, wv
L: 28. W: 71. Resembles Pacific Golden Plover but s*eparated by heavier, short-necked and pot-bellied appearance, comparatively shorter legs and bill, white underwing and voice; in flight, feet do not extend beyond tip of tail*. Flight pattern rather uniform with faint wing-bar mainly on primaries, no white in tail. Non-breeding and immature similar to Pacific Golden Plover, but yellow-buff supercilium less distinct, with *tertials falling well short of wing-tip*. Larger Grey Plover has black axillaries, stronger bill, white rump and grey appearance. **Voice** Barely disyllabic melancholy whistle *püyh* or repeated *pyü-pü* (often hard to place). **Habitat** Grassland, ploughed land, stubble, coast. **Note** Passage and winter hatched; scarce Oman, UAE; vagrant Bahrain, Kuwait, Saudi Arabia. [Alt: Eurasian Golden Plover]

American Golden Plover *Pluvialis dominica*   V
L: 26. W: 68. *Similar to Pacific Golden Plover* in size, structure and dark underwing. Separated from it in all plumages by *slightly shorter legs* (though still proportionately longer than European Golden, and toes project beyond tail-tip in flight) and *tertials falling far short of wing-tip and thus showing long primary projection* (extending well beyond tail). In Pacific Golden tertials longer, only slightly shorter than tail-tip, with short primary projection. In breeding plumage has darker upperparts (smaller yellow spots), *all-black underparts including flanks and vent* (whitish flank-line usually visible on ground and in flight in both European and Pacific Golden). Non-breeding and immature generally greyer than European and Pacific Golden with darker crown and upperparts, and *lacking yellow/greenish tone on neck and breast*; supercilium and forehead rather distinctly whitish (lacking yellow tone of Pacific); underparts grey-vermiculated (whitish belly in Pacific Golden). **Voice** Di- or trisyllabic, *similar to Pacific Golden*, but softer and more variable *kluilp, kuee-eep,* or shorter *hyyd*. **Habitat** Grasslands, wetlands, mudflats, sandy beaches. **Note** Vagrant Oman.

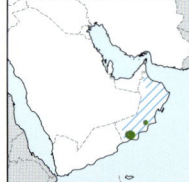

Pheasant-tailed Jacana *Hydrophasianus chirurgus*   WV, rb
L: 31 (48cm with full tail). Rail-like and often seen walking *on floating vegetation on very long toes*. Breeding adult has long, *black, downcurved tail*, chocolate-brown body, strikingly patterned head with *white face and foreneck, and golden-yellow hindneck, edged black. White wings with black tips conspicuous in flight*. In non-breeding plumage tail is short, *underparts turn white, but a dark breast-band remains, running up the neck and joining the eye-stripe*. Juvenile resembles non-breeding adult but breast-band flecked white, and head and neck pattern are duller. Low, rapid flight with dangling legs; landing with raised wings. **Habitat** Ponds, creeks and marshes with patches of open water and floating vegetation. **Note** Rare in hatched area; vagrant Qatar, Saudi Arabia, Yemen including Socotra.

PLATE 38: SMALL PLOVERS I

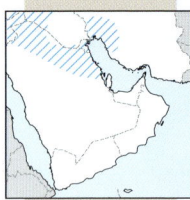

Eurasian Dotterel *Charadrius morinellus*   V

L: 21. W: 60. Brownish, dry-country plover; in non-breeding plumage recalls winter European Golden Plover, but smaller and with *whitish supercilia, meeting in 'V' on nape, and narrow, whitish upper breast-band* in all plumages. Adult breeding female (male duller) has striking *white supercilium and throat contrasting with blackish cap* and greyish neck, *upper breast bordered white, with chestnut and blackish below*. Adult non-breeding and immature with *brown, scaly upperparts and buff underparts. No wing-bar in flight*, but shows white tip to tail, and buffish-grey underwing; white shaft on outer primary sometimes rather distinctive. **Voice** Trilling, rather dry Dunlin-like *dryrrr*. **Habitat** Steppes and poor arable land. **Note** Passage and winter hatched; vagrant Bahrain, Oman, UAE.

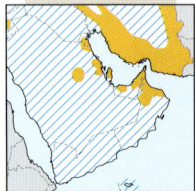

Little Ringed Plover *Charadrius dubius*   PM, WV, rb

L: 15. W: 45. Small, *slim, long-winged plover* with horizontal stance. In breeding plumage similar to Common Ringed Plover, but breast-band narrower and *lacks white wing-bar in flight*; note *yellow orbital ring, dark bill, white line behind black forecrown*, and muddy-coloured or pinkish legs. Adult non-breeding and juvenile have almost plain brown forehead with ill-defined pale patch behind or above eye and duller orbital ring; breast-band is broken or absent. Juvenile also has yellow-buff tinge to face and throat. *Separated from all other plovers by call and lack of wing-bar.* Mostly solitary or in pairs, sometimes small groups. **Voice** Flight and alarm call a loud, plaintive almost monosyllabic *diu*; in wavering display flight, with slow wingbeats, gives *pree-pree-pree* and tern-like *krre-u krre-u*. **Habitat** Mainly freshwater, in particular gravelly river islands and sandy borders of lakes; also coastal in winter. **Note** Passage hatched; some winter, mainly in S Arabia.

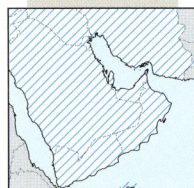

Common Ringed Plover *Charadrius hiaticula*   PM, WV

L: 19. W: 52. Small plover with *orange legs and black-tipped bill, black breast-band and white hindneck collar*. In flight, shows *conspicuous white wing-bar*. In non-breeding adult black is replaced by dark grey-brown, supercilium and forehead are tinged brown, and bill becomes all dark. Juvenile similar but is paler and duller, often with broken breast-band; upperparts with buff feather-fringes. Distinguished from Little Ringed Plover by white wing-bar, lack of pale orbital ring, and *pale supercilium in juvenile. Separated from other plovers by obvious white hindneck collar and call.* Mainly nominate subspecies occurs around Mediterranean basin, but further east most often only the subspecies *tundrae*, with darker upperparts, encountered. **Voice** A soft, rising, disyllabic whistle *tooip*. **Habitat** Sandy, muddy and stony shores, both coastal and inland. **Note** Passage and winter hatched; some oversummer.

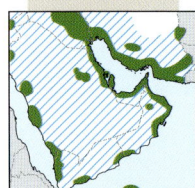

Kentish Plover *Charadrius alexandrinus*   RB, PM, WV

L: 16. W: 44. Small, sandy plover with white underparts, rather long, blackish legs, *conspicuous white hindneck collar and lacking complete breast-band*; in flight, a clear, white wing-bar and *broad white sides to tail*. Adult breeding male has variable *rufous cap, black frontal (forehead) bar, and lateral black breast-patches*. Breeding female, adult non-breeding and juvenile are duller, lacking black in plumage, and resemble non-breeding Common Ringed and Little Ringed Plovers, but separated by rather long, blackish legs, white breast, more white in tail, and call. By late-summer plumage of some individuals may bleach to entirely pale sandy-whitish above, with breast-patches also often entirely missing. **Voice** Flight call a soft *kip* or *twit*, recalling Little Stint; song a rattled repetition of *tjekke-tjekke* or *jid-id-jid... eer*. **Habitat** Shingle, sandy and muddy beaches, mudflats, mainly coastal but also inland, often on saline lagoons. **Note** Passage and winter hatched.

PLATE 39: SMALL PLOVERS II

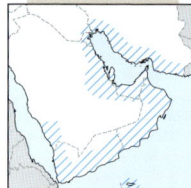

Lesser Sand Plover *Charadrius mongolus* — PM, WV

L: 20. W: 51. This species is split into two groups: the '*atrifrons* group' and the '*mongolus* group'; only *atrifrons* has been recorded in the region. Very similar to Greater Sand Plover in all plumages, behaviour and shape; lone individuals can be difficult to identify. Best separated by size (*body slightly larger than Common Ringed Plover; Greater Sand distinctly larger*); *shorter bill*, less pointed and with more swollen tip (but can overlap with Greater Sand); smaller, more rounded, less angular head; *shorter, darker, more greyish legs* (yellowish-green in Greater Sand). At rest, stance often more upright than Greater Sand. Breeding male has *all-black forehead and face mask*, *and broad reddish-chestnut breast-band* (Greater Sand never shows completely black forehead, and breast-band is usually narrower); female has black of head reduced (like Greater Sand). Birds of '*mongolus* group' very difficult to identify except in breeding plumage, when both male and female have prominent white forehead, and male often shows narrow, black margin to chestnut breast-band. Non-breeding and immature plumage (both groups) very like Greater Sand Plover, when structural features are important. *Separated from 'ringed' plovers by size and lack of white hindneck collar*; from Caspian Plover by shorter legs and wings, bolder white wing-bar, white underwing, and less bold supercilium. In flight, legs reach to or slightly beyond tip of tail (in Greater Sand legs show fairly prominently beyond); both sand plovers show clear, but variable, white wing-bar: in Lesser Sand Plover this is of more even width (in Greater Sand often most prominent on inner primaries). **Voice** Quieter than Greater Sand Plover, having a short, sharp and less trilling *chitik, chi-chi-chi, chik-tik*; also *kruu-kruit* or *drriiiit*. **Habitat** Tidal mudflats and sandy coasts. **Note** Passage and winter hatched, many oversummer; rare inland.

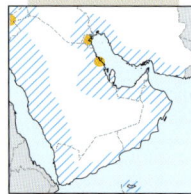

Greater Sand Plover *Charadrius leschenaultii* — PM, WV

L: 24. W: 56. Resembles larger version of Lesser Sand Plover, and lone individuals can be difficult to identify. Note especially the larger size (*obviously larger than Common Ringed Plover*); *longer, more pointed bill, larger, more angular head and large eye; longer yellowish-green legs, especially thighs (tibiae), with toes projecting well beyond tail-tip in flight*. Caspian Plover is slimmer with proportionately longer legs, and wings protrude well beyond tail when perched; head is more rounded with broader supercilium, underwing is dusky, and it has only a faint white wing-bar. Display flight recalls that of the ringed plovers, with dry chortling song incorporating Ruddy Turnstone-like call-note. **Voice** When flushed a trilling *kyrrr, kirr* or *trrr*; in song flight *huit-huit-huit* or ascending *dui-dui-tui-dit*. **Habitat** Breeds on inland sand- and mudflats, usually near water; otherwise mainly coastal. **Note** May breed Iran; passage and winter hatched. Some oversummer on Arabian coasts.

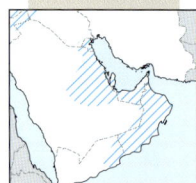

Caspian Plover *Charadrius asiaticus* — pm

L: 19. W: 58. *Slim, delicate plover* mostly recalling sand plovers in size and general appearance, but separated by *long, attenuated body with wings projecting well beyond tail-tip; proportionately longer legs*, long neck and smaller head with rather fine tapering bill; *broader white supercilium gives a capped appearance*. Male breeding shows distinct *blackish lower border to rufous breast-band*. In flight appears long-winged with faint wing-bar, only visible on inner primaries, and toes clearly project beyond tail; *underwings are dusky (not white)*, and tail is dark with less white at sides and tip. **Voice** Flight call is a short, sharp, *tyup*, sometimes repeated, and occasionally combined into a rapid series of rattling notes *tptptptptp*. **Habitat** Fields, grassy plains, semi-desert; also coastal areas. **Note** Generally a rare migrant, but regular in Arabia; vagrant Yemen.

PLATE 40: SNIPES AND WOODCOCK

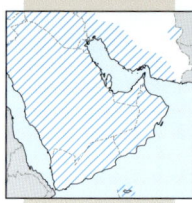

Jack Snipe *Lymnocryptes minimus*   pm, wv
L: 18 (incl. bill 4cm). W: 40. Very small with short bill. Well camouflaged and difficult to flush, usually rising silently at about one metre or less, in low, slightly jerky flight and settling almost immediately; two bright yellow lines on upperparts and short, pointed tail distinctive. On ground note green-glossed back, absence of central crown-stripe and dark-striped flanks (not barred). Separated from other snipes by size and short bill. Usually solitary. **Voice** Usually silent when flushed. **Habitat** Wetland margins. **Note** Passage and winter hatched, but mostly scarce.

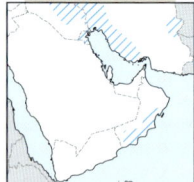

Great Snipe *Gallinago media*   pm, wv
L: 28 (incl. bill 6cm). W: 49. Larger, heavier and darker than Common and Pin-tailed Snipes with shorter bill and closely barred underparts and underwings. When flushed, rises at short distance, flies low and straight without zigzagging; flight heavy with rather slow wingbeats; pot-bellied appearance reminiscent of a small Woodcock. In flight, note absence of white belly-patch (shown by Common and Pin-tailed Snipes), dark underwing (similar to Pin-tailed, but unlike Common Snipe); white wing-bars on greater and median coverts, including primary coverts, more white on tail-corners and absence of white trailing edge to secondaries (conspicuous in Common Snipe). **Voice** When flushed low-pitched, rather weak, muffled *orrk*. **Habitat** Marshes, stubble fields, wet or dry rough grassland. **Note** Passage hatched, but rare; rare/vagrant Saudi Arabia; vagrant UAE, Yemen.

Common Snipe *Gallinago gallinago*   PM, WV
L: 26 (incl. bill 7cm). W: 45. Medium-sized with distinctly yellow-striped head and upperparts, dark-striped breast, barred flanks and very long bill; probes mud with vibrating movements. Often squats low; usually not seen until flushed at 10–15m distance; *rises explosively, immediately uttering several harsh calls, while zigzagging to a good height*. In flight, shows narrow *white trailing edge to wings* (lacking in Pin-tailed Snipe) and *white belly*. See Pin-tailed Snipe for further differences. **Voice** When flushed utters a few harsh *ärrtch* notes with slightly rising inflection. In aerial display produces distinctive reverberating sound ('drumming'). **Habitat** Wet grasslands, marshy water-margins. **Note** Passage and winter hatched.

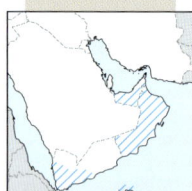

Pin-tailed Snipe *Gallinago stenura*   PM, WV
L: 26 (incl. bill 6.5cm). W: 45. Very similar to Common Snipe; best distinguished in flight, when note paler grey-buff panel on wing-coverts, *absence of white trailing edge to secondaries*, slightly more rounded wing-tips, and *completely barred underwing-coverts making underwing look dark greyish* (Common Snipe shows broad white bands, creating much paler underwing). On ground appears generally colder and darker above, with shorter, broader-based bill, shorter tail, *bulging supercilium in front of eye* (more parallel-sided in Common Snipe), *more scalloped scapular pattern* due to similar width and coloration of pale edges on inner and outer webs (in Common Snipe scapulars have broad white edges on outer webs only, contrasting with brown inner webs, creating white, often diagonal stripes) and *barred median wing-coverts* (more spotted in Common Snipe). When flushed, has slower take-off with slightly heavier, less erratic, flight; flies lower and drops after short distance. **Voice** A *short, sneezed dry* etch, often given just once; close to Common Snipe's call but shorter, without inflection. **Habitat** Wet fields, marshes, more often on drier ground than Common Snipe. **Note** Passage and winter hatched, but uncommon; vagrant Iran, Saudi Arabia.

Eurasian Woodcock *Scolopax rusticola*   V
L: 34. W: 58. Plumper and larger than the snipes; mainly in woodland. Usually seen when flushed; rises silently with zigzagging flight between trees, showing long bill, round body, broad wings, red-brown rump and tail with black subterminal band. On ground note large eyes set far back, barred crown (striped in snipes), and completely barred buff underparts. Often only flushes almost underfoot. **Voice** Displays at dusk in level, direct flight low over treetops, with distinctive call: one or more deep croaks followed by sharp *twzzip*. **Habitat** Woodland floor, scrub, thick cover. **Note** Winter hatched; vagrant Bahrain, Oman, Saudi Arabia, UAE.

PLATE 41: DOWITCHERS AND GODWITS

Long-billed Dowitcher *Limnodromus scolopaceus* — pm, wv

L: 28. W: 49. Plumage and structure *resemble small Bar-tailed Godwit with straight, snipe-like bill, slightly drooping at tip*; legs pale olive-green. *In breeding plumage underparts orange-red*, densely barred or spotted blackish; upperparts brown, narrowly fringed rufous. Non-breeding/first-winter has pale grey underparts with darker neck and breast and slightly barred flanks and undertail-coverts. In all plumages *distinctive pale supercilium contrasts with black lores and dark crown*. In flight has *narrow white oval on back and narrow white trailing edge to secondaries*. When flushed, rises fast, snipe-like, but with slower wingbeats. **Voice** Sharp, clear, sometimes repeated, *keek*, slightly reminiscent of Oystercatcher. **Habitat** Muddy freshwater pools. **Note** Vagrant Oman.

Asian Dowitcher *Limnodromus semipalmatus* — V

L: 35. Easily confused with Bar-tailed Godwit, with which it may associate. Told by *smaller size and straight, all-black bill, slightly swollen at tip* (Bar-tailed noticeably pink-based, pointed and slightly upturned), and *held more angled below horizontal* (about 30°; only 10–15° below in Bar-tailed). *Feeding action with continuous vertical 'sewing-machine' probing*. Summer plumage similar to Bar-tailed, but note *whitish ventral region and white flanks, barred greyish and chestnut* (Bar-tailed lacks darker-barred flanks, males having wholly rusty-red underparts). In grey-and-white winter plumage has slightly mottled grey-brown breast, obscurely barred flanks, and bolder eye-stripe than Bar-tailed; in juvenile plumage, neck and breast have stronger buff wash than juvenile Bar-tailed. *In flight, white rump is barred dark, contrasting only slightly with back and tail, unlike Bar-tailed*. See also Long-billed Dowitcher. **Voice** Yelping *chep-chep* or *chowp*; also *aow*, recalling distant human cry. **Habitat** Mudflats, sandbanks. **Note** Vagrant Oman, Yemen.

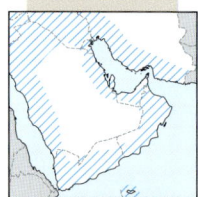

Black-tailed Godwit *Limosa limosa* — PM, WV

L: 42. W: 77. Large with long bill and legs. Similar to Bar-tailed Godwit, but slightly larger and more erect due to *longer legs and neck; bill slightly longer and straighter and totally different upperwing pattern* (also visible from below). In summer plumage, both sexes have varying amount of *rusty-orange on head, neck and fore-body* with *diffuse, dark bars on lower breast, fore-belly and flanks*. Female duller, sometimes predominantly greyish. Non-breeding plumage uniform grey, washed ochre on neck and breast in juvenile, and with pink-based bill. In all plumages easily *distinguished in flight by broad, white wing-bar, white tail with black terminal band and trailing legs and feet*. Often in flocks, including when feeding. **Voice** All calls rather nasal and scolding; alarm call *titi-teev* or, from more excited birds, *wicka-wicka-wicka*. **Habitat** Muddy freshwater margins, grassland, marshes, estuaries, tidal creeks. **Note** Passage and winter hatched, some oversummer in Arabia.

Bar-tailed Godwit *Limosa lapponica* — PM, WV

L: 38. W: 75. Resembles Black-tailed Godwit, especially on ground in non-breeding plumage, but note more compact, less erect appearance with *shorter legs and neck*, and *slightly upturned bill*. Easily separated from Black-tailed in flight by lack of obvious wing-bar (and resembling Whimbrel). In breeding plumage has *deep rusty-red head and entire underparts*; larger female is buffish or faintly rusty, *lacking breast- or belly-bars of Black-tailed Godwit*. Non-breeding/winter plumage buffish-grey with dark shaft-streaks above and on breast (almost uniform smooth grey in Black-tailed). Juvenile similar but more buffish with darker upperparts lacking shaft-streaks and breast markings. Gregarious at roosts but does not feed in flocks. **Voice** In flight, or when flushed, soft, low-pitched, nasal *beb-beb*, or sharper *ke-kek*. Often silent. **Habitat** Mudflats, sandy beaches, estuaries. **Note** Passage and winter hatched; many oversummer on coasts.

Long-billed Dowitcher

winter

Asian Dowitcher

juv moulting to 1st-winter

ad summer

Black-tailed Godwit

ad summer

winter

winter

♂ summer

Bar-tailed Godwit

winter

PLATE 42: WHIMBREL AND CURLEWS

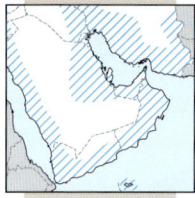

Whimbrel *Numenius phaeopus* — PM, WV
L: 41. W: 83. *Smaller and slightly darker than Eurasian Curlew* with faster wingbeats. *Bill usually shorter and more decurved near the tip* (though overlaps with young Eurasian Curlew in length) and head shows *dark crown with pale central stripe and dark eye-stripe*. Similar in flight to Eurasian Curlew; body and upperwing usually appearing darker than in Curlew. **Voice** Flight call *characteristic, fast series* of whistled notes, a tooted *bi-bi-bi-bi-bi-bi*, quite unlike the soft rising whistle of Eurasian Curlew. **Habitat** Estuaries, sandy beaches, rocky shores, coral reefs, grasslands. **Note** Passage hatched, rare inland; winters Iranian and Arabian coasts, where some also summer.

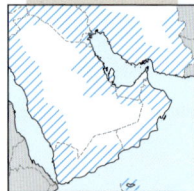

Eurasian Curlew *Numenius arquata* — PM, WV
L: 55. W: 90. Large, streaked, brownish wader with *long, decurved bill*. Male smaller than female, with shorter bill, which in young male overlaps with Whimbrel, but is more evenly decurved. Flight with rather slow, gull-like wingbeats, showing barred tail and white wedge on rump and lower back (similar to Whimbrel), but *feet projec*t, unlike in Whimbrel. Otherwise told from Whimbrel by size, *usually longer bill, uniform head pattern and flight call*. The subspecies *orientalis*, the most frequent in Arabia, has white underwing and much longer bill – not to be confused with Far Eastern Curlew, which has brownish underparts and rump and much darker underwing. **Voice** Flight call is a drawn-out, melodic, slowly rising whistle, easily imitated, *cour-leee*. **Habitat** Tidal mudflats and sands, rocky shores, coral reefs; also inland on muddy or grassy wetland margins. **Note** Passage and winter hatched; some summer in Arabia.

Slender-billed Curlew *Numenius tenuirostris* — V
L: 39. W: 84. *Much smaller and slimmer than Eurasian Curlew* (smaller even than Whimbrel) with slim neck – never showing the 'camel-shaped' neck of Eurasian Curlew. *Shorter, finer, all-dark bill tapering to a thin tip* (may show slightly paler base to lower mandible). Plumage 'cleaner' with slightly *greyer upperparts and whiter, more contrasting breast and underparts, showing black, rounded spots on flanks and sides of belly in adult* (brown streaks in juvenile); usually has darker crown (slightly capped appearance) and lores, and paler supercilium than Eurasian Curlew. In flight, *dark outer primaries and primary coverts contrast more with paler inner wing than in Whimbrel and most Eurasian Curlews*, but *underwing pure white* (darker in Whimbrel and many Eurasian Curlews except paler, eastern *orientalis*). Often adopts an upright stance and moves more rapidly on ground than Eurasian Curlew or Whimbrel. **Voice** Flight call most similar to Eurasian Curlew but shorter and higher pitched, lacking the liquid quality, a single *cour-ee*. **Habitat** Steppe and semi-desert fringes of freshwater wetlands. **Note** Former vagrant Oman, Yemen, possibly Saudi Arabia; may now be extinct.

Far Eastern Curlew *Numenius madagascariensis* — V
L: 63. Resembles Eurasian Curlew, but separated in flight by *all-brown upperparts (thus lacking white wedge up lower back)*; underwing and axillaries completely barred brownish forming *dark underwing* (almost white underwing-coverts in Eurasian Curlew). Plumage generally darker, browner than Eurasian Curlew, with dark buff head, neck and all underparts (though slightly paler rear belly in winter); *very long bill* (but overlaps in length with that of *orientalis* Eurasian Curlew). **Voice** Flight call similar to Eurasian Curlew, but less fluty *krr-iii*, more clearly disyllabic, with first note rather harsh and second note longer; higher pitched and lacking rising inflection. **Habitat** Marshes, sea coasts, mudflats. **Note** Vagrant Iran, Oman. [Alt: Eastern Curlew]

PLATE 43: SANDPIPERS I

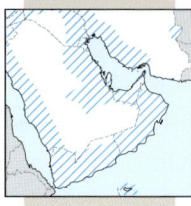

Spotted Redshank *Tringa erythropus* — PM, WV
L: 30. W: 64. Medium-sized, *rather slim wader with red legs*. In winter plumage recalls Common Redshank, but larger with paler grey upperparts, white underparts, and *longer, finer bill*, slightly drooping at tip. *Upper mandible wholly black*, lower red with black tip. In breeding plumage *black, finely spotted white*. In flight, shows all-dark wings, *distinctive white rump extending in wedge up back*, and feet projecting well beyond tail. **Voice** Flight call distinct piercing disyllabic *tju-it*. **Habitat** As Common Redshank, but freshwater and lagoons more than open tidal settings. **Note** Passage and winter hatched.

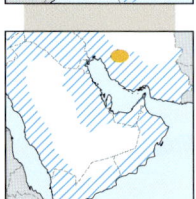

Common Redshank *Tringa totanus* — PM, WV
L: 28. W: 62. Greyish with *bright red legs*, in flight showing *broad white trailing edge to wings and white on rump extending up back*. Breeding adult dark-spotted, with red base to bill. Winter plumage uniform grey above, paler grey below with ill-defined spotted breast. Juvenile has buff-spotted upperparts and all-dark bill. Larger and slimmer juvenile Spotted Redshank has greyer upperparts, heavily barred or vermiculated underparts, finer, longer bill and different wing pattern. **Voice** Calls distinctive; disyllabic *djü-dü*, with stress on first syllable; alarm a persistent *tjü-tjü-tjü....* **Habitat** Breeds in damp grassland; on passage and winter mainly coastal shores, mudflats; also inland wetlands. **Note** Passage and winter hatched.

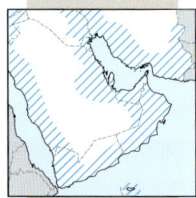

Marsh Sandpiper *Tringa stagnatilis* — PM, WV
L: 23. W: 57. *Resembles small, slim Common Greenshank with proportionately longer legs and thin, straight, finely pointed bill. Face almost white, usually with distinctive supercilium. Flight pattern similar to Greenshank with dark wings, contrasting white tail and wedge up back; protruding legs distinctive beyond tip of tail.* Flight action rapid and more similar to Wood Sandpiper. Winter plumage rather uniform, paler grey than Greenshank with almost white face and distinctive supercilium. In breeding plumage, becomes markedly black-spotted on head and upperparts; legs usually more yellowish. **Voice** Flight call a clear *djeeu-djeeu* often repeated, weaker and less shrill than Common Greenshank. **Habitat** Freshwater wetlands, tidal creeks and flats. **Note** Passage and winter hatched.

Common Greenshank *Tringa nebularia* — PM, WV
L: 32. W: 69. *Larger than Common Redshank with long greenish legs and fairly long, slightly upturned greyish-green bill.* Flight action slow and jerky with long, *dark wings contrasting with paler head and neck, and conspicuous white tail and wedge up back*. In winter plumage resembles Marsh Sandpiper but darker grey, and *darker face lacks pale supercilium*. Summer plumage has black feather-spotting on upperparts and distinctly spotted breast and flanks. Juvenile uniformly patterned with buff-fringed grey-brown upperparts, white underparts and streaked grey head, neck and breast. Rather active, often running when feeding in shallow water. **Voice** Flight call characteristic shrill trisyllabic *djiu-djiu-djiu*, with equal stress on all syllables. Marsh Sandpiper's call is similar but thinner and less shrill. **Habitat** Coastal shores, mudflats, inland wetlands. **Note** Passage and winter hatched; some oversummer Oman, UAE.

Lesser Yellowlegs *Tringa flavipes* — V
L: 23–25. *Marginally smaller than Common Redshank and clearly more slender with long primary projection. Legs long, bright yellow*. Plumage finely spotted above, with breast diffusely streaked. In flight wing and tail pattern, especially square white rump, recalls Wood Sandpiper. Bill fine, straight and all dark. Legs may be paler yellow, even orangey, in winter or in young birds. Supercilium short, reaching only to eye (unlike Wood Sandpiper). **Voice** Clear *tew*, close to that of Common Greenshank and Marsh Sandpiper. **Habitat** Mainly freshwater edges. **Note** Vagrant Oman, UAE.

PLATE 44: SANDPIPERS II

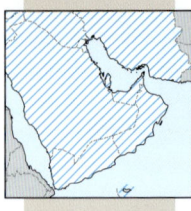

Green Sandpiper *Tringa ochropus* — PM, WV
L: 23. W: 59. Distinguished by call and contrasting black-and-white plumage. Shy, often first seen when flushed, when *black wings above and below contrast sharply with white belly and white rump*. At rest blackish upperparts and breast contrast with clear white belly and flanks. Juvenile darker, more uniform and buff-spotted, head rather dark with obvious pale eye-ring and short supercilium in front of eye. Often in small flocks. **Voice** Flight call sharp, melodic *dlo-eed-witt-witt*, loud and far-carrying. **Habitat** Muddy streams, small pools, wadis, edges of freshwater wetlands. **Note** Passage and winter hatched.

Wood Sandpiper *Tringa glareola* — PM, WV
L: 20. W: 56. Resembles Green Sandpiper but *upperparts paler, less contrasting and boldly speckled whitish and with conspicuous white supercilium*; also rather long, yellowish-green legs and more elegant appearance. In flight brownish upperparts contrast with white rump; di*ffers from Green Sandpiper in whitish (not dark) underwing, longer feet-projection* and call. Juvenile has buff-spotted upperparts, somewhat recalling young Common Redshank. Occurs singly or in small flocks. **Voice** When flushed, characteristic, far-carrying series *jiff-iff-iff-iff*. **Habitat** Freshwater marshes with muddy margins. **Note** Passage and winter hatched.

Terek Sandpiper *Xenus cinereus* — PM, WV
L: 23. W: 58. Rather squat, front-heavy wader with short *orange-yellow legs and long upcurved bill*. Grey head and upperparts *bordered by dark shoulder bar, carpal patch and primary-line*. In flight, *wings show white trailing edge*, much narrower than that of Common Redshank. In juvenile and winter adult, black shoulder bar is faint or absent. *Frequently bobs rear end* (reminiscent of Common Sandpiper) and when feeding often runs fast with head lowered. Feeds singly, but may roost in groups. **Voice** Flight call clear, fluty *tjiy-tjiy* or *dwitt-dwitt*, softer than Common Redshank, sometimes recalling Ruddy Turnstone in character. **Habitat** Tidal mudflats, saltmarshes, mangrove creeks, coral reefs; scarce inland on passage. **Note** Passage hatched; winters Arabian coasts (where some oversummer).

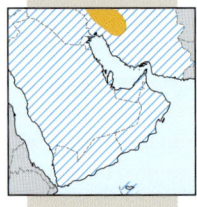

Common Sandpiper *Actitis hypoleucos* — PM, WV
L: 20. W: 40. Short-legged wader with almost uniform brown upperparts, clean white underparts running up in a wedge between wing and brownish breast-sides; rather long tail projecting well beyond wing-tips at rest; note constantly bobbing rear-body. Unique flight action low over water with vibrating, shallow and stiff wingbeats, alternating with short glides. In flight shows distinct, but rather thin, white wing-bar (also obvious from below) and brown rump and tail with pale outer edges. Juvenile has wing-coverts barred buff and dark. Feeds singly, sometimes roosts in small groups. **Voice** Flight call characteristic series of piping notes, descending *hee-dee-dee-dee*. **Habitat** Winter and passage on edge of any wetland. **Note** Passage and winter hatched.

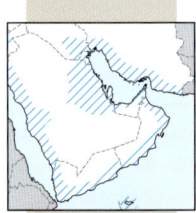

Ruddy Turnstone *Arenaria interpres* — PM, WV
L: 23. W: 53. Stocky, *short-billed and short-legged wader*, distinguished in breeding plumage by *striking head and breast markings, chestnut shoulders and wing-coverts, and bright orange legs*. Flight is strong and direct and shows distinctive pied and chestnut appearance; *black breast is a striking feature*. In winter and juvenile plumage, head and neck are much darker, and usually lack chestnut; also juvenile shows bold buffish fringes on most coverts. Walks with a rolling gait, and turns or flips over stones and seaweed for food items. **Voice** Calls short and staccato; when feeding low *chuk*; in flight *krytt-te- krytt-te-krytt* or *kritt-it-it*. **Habitat** Rocky or sandy coasts; rarely inland lakes. **Note** Passage and winter hatched; occasional inland. Some remain in summer.

PLATE 45: SANDPIPERS III

Little Stint *Calidris minuta*   PM, WV
L: 13. W: 30. Abundant in region. *Almost always has dark bill and legs*, the latter separating it from Temminck's Stint. In summer plumage, colour on face, neck, breast and scapulars varies from dull orange to warm buff; but *always shows pale 'V' on mantle and dark centre to crown*. In winter, upperparts become grey, usually with dark shaft-streaks to feathers. Juvenile, which looks very white below, has warm rufous tone to upperparts, a distinct white 'V' on mantle, diffuse greyish neck-collar and rufous, streaked breast-sides. Feeding action rapid. **Voice** Flight call a short *tip*, *chit* or trill. **Habitat** Coastal flats and inland wetlands. **Note** Passage and winter hatched; some oversummer in Arabia.

Temminck's Stint *Calidris temminckii*   PM, WV
L: 13. W: 30. Similar in size to Little Stint but with more elongated body and *shorter, yellowish-green legs*; *white sides to tail prominent in flight*. In summer, mainly grey-buff with *dark centres to many scapulars*, and lacking rufous-orange tones. In winter the dark scapulars are lost and then looks plain buff-grey, but note defined grey breast. Juveniles have narrow buffish fringes to scapulars and coverts with some dark markings on upper scapulars. *Lacks white 'V' on mantle in all plumages*. Often keeps tipped forward, legs flexed. Sometimes in small groups; often towers high in erratic flight when flushed. **Voice** Flight call distinctive ringing trill, *tirrrrr*, quite unlike Little Stint. **Habitat** Inland pools, marshes, muddy coasts. **Note** Passage hatched; winters mainly S Iran and Arabia.

Long-toed Stint *Calidris subminuta*   pm, wv
L: 14. W: 30. Similar to Little Stint in size but with *longer neck* (noticeable when standing upright) and *longer legs, which are dull yellowish or yellowish-brown*; the long toes can be difficult to see. Leg colour similar to Temminck's Stint, from which told by longer legs, more upright posture, and plumage. In summer, rufous and well-streaked, with noticeable supercilium creating capped appearance. In winter, note dark feather centres to upperparts and fine streaking on head and breast, unlike the plain grey of Little and Temminck's Stints. When flushed, shows faint wing-bar and often towers high like Temminck's Stint. **Voice** Short, soft *prrt* or *tit-tit-tit*, in flight. **Habitat** Freshwater margins, coastal pools, mudflats. **Note** Passage and winter hatched, but scarce; vagrant Bahrain, Iran, Saudi Arabia, Yemen.

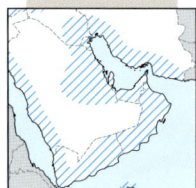

Sanderling *Calidris alba*   PM, WV
L: 20. W: 40. Slightly larger than Dunlin with *shorter, straighter bill*. In winter has *very pale plumage with dark mark at bend of wing*. In summer and juvenile plumage more easily confusable with other small waders, particularly stints. Note *lack of hind toe*. In summer plumage can be quite *rusty on head and has prominent breast-band, but always shows dark scalloping in the red of the breast*, a feature that helps separate from smaller, vagrant Red-necked Stint. Juvenile *spangled black and white on upperparts*, sometimes with buff wash on breast-sides. *White wing-bar noticeable in flight*. *Often runs fast*, particularly on open shoreline ahead of wave-wash. **Voice** Usual flight call is a loud *plit*. **Habitat** Sandy beaches, mudflats. **Note** Passage hatched, rare inland; coastal in winter.

PLATE 46: SANDPIPERS IV

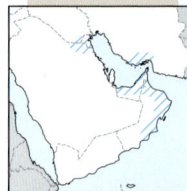

Great Knot *Calidris tenuirostris*   pm, wv
L: 27. W: 55. Like large Red Knot, similarly stocky build but more tapering at rear. In winter plumage told from Red Knot by *longer, slightly more decurved bill with heavier base, larger greyish spots on underparts, less defined supercilium (more extensive greyish lores) and more obvious tail pattern (white rump and uppertail-coverts contrasting with dark upperparts and tail)*; upperparts darker than Red Knot with dark streaking to centres of grey feathers. In breeding plumage easily told by *dense blackish spotting on breast* and flanks, dark streaks on mantle and hindneck, and *chestnut centres to scapulars*. Juvenile also told from similar Red Knot by *heavily marked breast, contrasting with pale belly, and pale-fringed wing-coverts with dark shaft-streaks*. In first-winter some juvenile coverts often retained, helping identification. **Voice** Usually silent, occasionally a soft *prrt*. **Habitat** Coastal mudflats. **Note** Passage and winter hatched, but often rare; vagrant Bahrain, Qatar, Saudi Arabia, Yemen.

Red Knot *Calidris canutus*   V
L: 24. W: 50. In winter told by combination of stocky build (noticeably larger and plumper than Dunlin), pale grey upperparts and straight, rather stout bill (about length of head); in flight, rather long-winged, rump greyish-white and tail grey. Legs grey-green. In summer plumage, brick red below with black, white, grey and buff mottling above; told from summer Curlew Sandpiper by shorter straight bill, larger size and tail pattern. Juvenile has poorly marked breast (unlike Great Knot); wing-coverts have dark subterminal markings and pale tips, but no dark shaft-streaks as in Great Knot. **Voice** Short nasal *wut* or *wut wut*, rather quiet. **Habitat** Coastal mudflats. **Note** Vagrant Near East, Iraq, Iran and Arabia.

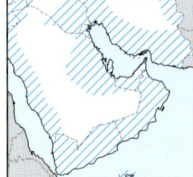

Curlew Sandpiper *Calidris ferruginea*   PM, WV
L: 19. W: 40. Told from Dunlin in all plumages by *white rump, longer, more decurved bill and longer legs*, giving a more elegant appearance. Easily told in breeding plumage by *chestnut-red face and underparts* (often with white feather-fringes). In winter plain grey above, white below with a light suffusion to breast-sides and *noticeable white supercilium*; then also distinguished from Dunlin by cleaner appearance with whiter underparts. In first-autumn, has rather scaly grey-brown upperparts, noticeable white supercilium and yellowish-buff wash to breast. Usually in small flocks. **Voice** Trilling, *trururip* or almost disyllabic *churrip* in flight. **Habitat** Coastal mudflats; also inland wetlands **Note** Passage hatched; winters mainly in coastal Arabia (some oversummer).

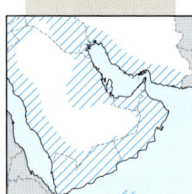

Dunlin *Calidris alpina*   PM, WV
L: 18. W: 38. Larger than the stints with *longer, slightly downcurved bill*. In summer plumage easily told by *black belly-patch*. In winter, this patch is lost as are rufous tones to upperparts; then has grey upperparts with narrow pale fringes to coverts and scapulars, grey breast and white belly. From similar Curlew Sandpiper by different rump pattern (*white with dark centre*), lack of white supercilium, darker upperparts and breast, and shorter bill and legs. In first-autumn, note chestnut on coverts, white 'V' on mantle, and lines of dark spotting on flanks below finely streaked breast. **Voice** Reedy *kreep* in flight distinctive. **Habitat** Coastal mudflats; also inland wetlands. **Note** Passage and winter hatched; some summer south to Arabia.

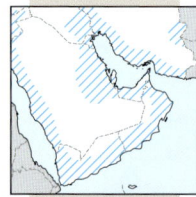

Broad-billed Sandpiper *Calidris falcinellus*   PM, WV
L: 17. W: 35. Slightly smaller than Dunlin, from which told by *longer, broad-based bill with downward droop near tip* (head-on, tip also seen to be swollen), *shorter yellowish-grey legs* and *double supercilium*. In summer upperparts rather dark with white 'V' on mantle, white line on scapulars and *white underparts with dark-spotted and streaked breast and flanks*. In winter, greyer with less obvious supercilium; then shape and length of bill, leg colour and, if present, dark area on carpals important for identification. In first-autumn, resembles adult but streaking below finer and confined to breast. Often has slow-moving and crouching feeding action. In rather erratic flight, appears small, heavy-fronted and, in breeding plumage, dark with narrow wing-bar. **Voice** Flight call rather weak with a dry, slightly buzzing character, ascending a little at end, *brlliid*. **Habitat** Coastal mudflats; also inland wetlands. **Note** Passage hatched; winters mainly in coastal Arabia. Formerly placed in genus *Limicola*.

PLATE 47: RARE SANDPIPERS AND PAINTED-SNIPE

Baird's Sandpiper *Calidris bairdii* V
L: 15. W: 38. Fairly small sandpiper with distinctive shape *created by long wing projection beyond tip of tail* and rather short legs. Most likely to occur in region in juvenile plumage, when told by shape, *scaly upperparts* (white fringes to grey-buff feathers), faintly streaked breast on buffish ground colour, pale spot above lores (at close range), dark legs and bill. In flight, looks long-winged with poorly marked wing-bar and poorly defined tail/rump pattern. **Voice** Short, rolling *krru*. **Habitat** Sandy shores, lagoons, wetland margins. **Note** Vagrant Oman.

Pectoral Sandpiper *Calidris melanotos* pm
L: 21. W: 42. Long-winged, brown sandpiper, *larger than Dunlin* with short, slightly decurved bill and, when not feeding, often with rather long-necked appearance. Told in all plumages by *clearly demarcated streaked breast contrasting with white belly, dull yellowish legs and faint wing-bar*. **Voice** Loud *kreet* in flight. **Habitat** Marshes, pools with grassy edges. **Note** Rare Oman; Vagrant UAE.

Sharp-tailed Sandpiper *Calidris acuminata* V
L: 20, W. 43. Similar to Pectoral Sandpiper in structure and plumage but with longer legs and shorter bill. Differs in all plumages by *lack of sharp demarcation between streaked breast and white belly, white supercilium accentuating rufous crown* and, at close range, white eye-ring. Adult in summer has dark arrow-shaped marks on underparts, which are lost in winter to give finely grey-streaked breast-sides. In winter, upperparts lack any rufous, becoming grey with darker feather-centres. In first autumn, bright rufous-and-buff-streaked upperparts with whitish lines on mantle and scapulars; upper breast with noticeable pale orange wash. **Voice** Flight call short, soft, metallic *pleep*, often rapidly repeated. **Habitat** Marshes, pools with grassy edges. **Note** Vagrant Oman, Yemen.

Buff-breasted Sandpiper *Calidris subruficollis* V
L: 19. W: 45. Slender, buffish Ruff-like wader with small, rounded head with buff face and prominent black eye. *Short, straight bill and bright yellow-ochre legs; upperparts distinctly scaly, and underparts from head to undertail buff with fine spotting on breast-sides*. In flight, appears rather plain with *no white in wings or tail*. Juvenile similar to adult. From young female Ruff by smaller size, shorter straight bill, rounded head, clean buff face, spotted breast-sides, brighter yellowish legs and different upperpart pattern in flight. **Habitat** Fields, short grass, in preference to shores. **Note** Vagrant Oman, Saudi Arabia, UAE. Formerly placed in genus *Tryngites*.

Greater Painted-snipe *Rostratula benghalensis* V
L: 25. W: 52. Skulking; when flushed note dangling legs, short tail, rounded wings and rather slow flight. *Bill pale, shorter than Common Snipe and slightly decurved at tip.* Distinctive head markings, buff 'V' on mantle and *white line in front of wing*; dark foreparts contrast sharply with white belly. *Male has prominent buff eye-ring and streak behind eye.* In flight, shows golden-buff spotted barring on forewing and flight feathers, and black wing-bar; also *Ruff-like white rump pattern. Larger and brighter female mainly greeny-bronze* with *chestnut head and neck with distinctive white marks* (similar to male). **Habitat** Marshes with muddy patches, reedbeds; will feed on open fields. **Note** Vagrant Iran, Oman, SW Saudi Arabia, UAE.

PLATE 48: RUFF AND PHALAROPES

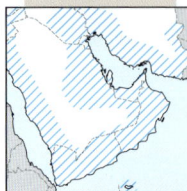

Ruff *Calidris pugnax* — PM, WV
L: 28 (male); 22 (female). W: 56 (male). Male about size of Common Redshank, but more upright with *longer neck, smaller head* and proportionately *shorter, slightly drooping bill*; often looks humpbacked and pot-bellied. Flight lazy, sometimes interrupted with glides; shows narrow wing-bar and characteristic *oval white patches at tail-base*. Adult in summer plumage variable: upperparts, breast and flanks show mix of black, brown, chestnut, ochre and white, heavily barred or blotched (male may show large ruffs and ear-tufts in spring). Plumage greyer in winter, when *lores always pale*, face often whitish and rear belly and undertail always white; a few males show white on head and breast. Bill mostly blackish-brown, but can be yellow or pinkish, tipped dark; *legs vary from orange-red to greenish-grey*. In juvenile, upperparts scaly, blackish-brown distinctly fringed buff, and head, neck and breast yellowish-brown, tinged orange; legs are yellowish-brown or greenish. Often in loose flocks. **Habitat** Inland wetland fringes, wet grassland, coastal lagoons. **Note** Passage and winter hatched. Formerly placed in genus *Philomachus*.

Wilson's Phalarope *Phalaropus tricolor* — V
L: 23. W: 41. Resembles Marsh Sandpiper more than the other two phalaropes, but has shorter, brighter legs, *long needle-like black bill, and white rump* (not white wedge extending up back). Feeds as much on land as in water and very active. In flight shows *white rump, plain greyish upperwing without wing-bar* and feet projecting beyond tip of tail. In winter plumage, mainly *grey with white underparts and yellow legs*. In breeding plumage, albeit unlikely to be seen in region, has a colourful and striking head and neck pattern, especially in female. **Habitat** Fresh and brackish waters. **Note** Vagrant Oman, UAE.

Red-necked Phalarope *Phalaropus lobatus* — PM, WV
L: 18. W: 36. Small, elegant wader *which swims* high on water, often spinning to whirl up food items, which will be taken with *needle-thin, black bill*. In breeding plumage has striking head and neck pattern (female is brighter than male). In winter plumage upperparts are pale grey with *black mask through eye and black on hindcrown* conspicuous. Juvenile has similar head pattern, but *upperparts are dark brown with prominent ochre bands*. Flight fast and jerky, wings show distinct white wing-bar; *often suddenly settles on water*. Separated from Grey Phalarope in all plumages by smaller size and very thin black bill; in winter, Grey Phalarope is also paler and more uniform above. Gregarious; often in large flocks at sea. **Voice** Flight call short, sharp *kritt* or *kitt* recalling Sanderling, but quieter and finer. **Habitat** Maritime; inland lakes on passage. **Note** Passage hatched, winters in Arabian Sea.

Red Phalarope *Phalaropus fulicarius* — pm, wv
L: 21. W: 42. Slightly larger and more robust than Red-necked Phalarope, with shorter, thicker neck; *bill distinctly thicker, less pointed and sometimes pale (yellowish) at base*. Red breeding plumage with black-and-white head pattern is not likely to be seen in the region. Winter plumage recalls a very small gull when feeding on water with *unmarked grey upperparts* (paler, more uniform than Red-necked). First-winter similar to adult, but black on hindneck sometimes remains in winter. Behaviour and flight pattern similar to Red-necked Phalarope, but wingbeats less jerky. **Voice** Flight call a sharp *pik*. **Habitat** Essentially maritime, but vagrants may visit lakes, pools or puddles, large or small. **Note** Widespread vagrant in Middle East. [Alt: Grey Phalarope]

PLATE 49: COURSER AND PRATINCOLES

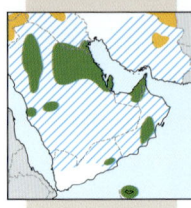

Cream-coloured Courser *Cursorius cursor* — PM, WV, rb
L: 24. W: 54. *Sandy-buff with distinctive black-and-white head markings joining in 'V' on nape; bill short and downcurved.* Erect posture; runs quickly with plover-like behaviour and sudden stops; prefers running away rather than flying. Flight rapid with long, slightly rounded wings and legs protruding well beyond tail. Wing pattern distinctive, *black outer wing above and black underwing contrast sharply with rest of plumage.* Juvenile lacks grey and black on crown; upperparts, head and breast have faint brown spots or irregular dark subterminal lines, and primaries are fringed buff. Sometimes in post-breeding flocks. **Voice** Flight call is a short *kwit-kwit.* **Habitat** Sandy or stony semi-desert with scanty vegetation, marginal cultivation, arid flat country; post-breeding, also short grassland, fields and turf farms. **Note** Post-breeding dispersal and passage hatched.

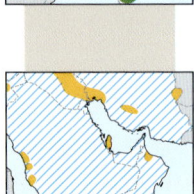

Collared Pratincole *Glareola pratincola* — PM, mb
L: 25. W: 63. Distinctive, aerial wader, resembling marsh terns, having *graceful fast flight, long pointed wings, deeply forked tail and short bill.* Usually seen in loose flocks chasing winged insects. On ground, plover-like with quick tripping actions on short legs, often with upright stance, head held high. *Adult has creamy-buff throat bordered black*; tail and wing-tips are equal in length. In flight, uniform dark olive-brown above (slightly darker flight feathers) with *narrow, but distinctly white trailing edge to secondaries*, contrasting with white rump and belly; *underwing-coverts reddish-brown*, but often look shadowy black. *Note: some, juveniles/worn adults may lack or have only thin white trailing edge to wings.* (Black-winged Pratincole has generally darker upperparts, lacks white trailing edge to wings, and has black underwing-coverts.) Adult non-breeding and juvenile lack distinct black throat line, and juvenile's brown feathers have pale tips and fringes; outer tail feathers are shorter than in adult. **Voice** Most characteristic call is a tern-like, sharp, chattering *kikki-kirrik* and a short *check* or *che-keck*. **Habitat** Sun-baked mudflats and flat, firm plains with low vegetation, grasslands; often near water. Nests colonially. **Note** Has bred in N Oman; passage hatched, a few winter in S Arabia.

Oriental Pratincole *Glareola maldivarum* — No records
L: 23. W: 63. Difficult to identify. Has distinctive features of, and resembles, Collared and Black-winged Pratincoles closely. Colour of upperparts and *lack of white trailing edge to secondaries (a thin pale fringe may show in fresh-plumaged adult or juveniles) are similar to Black-winged, but shares reddish-brown underwing-coverts with Collared Pratincole.* Told from both in adult plumage by *obviously shorter tail, the tip of which falls short of wing-tip*, being best judged when settled, but shortness of tail is certainly noticeable in flight. Adult has underside of outermost tail feather with restricted black tip (distal half black in Collared). Combination of features as given here may sometimes be required for conclusive identification; beware active or suspended moult, or damaged tail, in Collared Pratincole may produce apparently 'short-tailed' individuals. **Habitat** As Collared Pratincole. **Note** Vagrant Iran, UAE. No accepted records for Oman, but a potential vagrant.

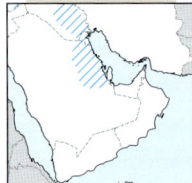

Black-winged Pratincole *Glareola nordmanni* — V
L: 24. W: 64. Structure and behaviour as Collared Pratincole and often difficult to distinguish between the two. Adult slightly darker above than Collared, with no contrast between flight feathers and coverts; *lacks white trailing edge to secondaries, and underwing-coverts are jet-black* (although these coverts are reddish-brown in Collared Pratincole this can be hard to assess, and they often appear dark); tail fork of Black-winged shallower. When perched, wings project beyond tail-tip (in Collared approximately equal), and shows less red at base of bill. Non-breeding adult and juvenile resemble corresponding Collared Pratincole but identifiable in flight by wing pattern. **Voice** Call resembles Collared, though a little sharper, and in breeding area alarm call is a shorter *pwik* or *pwik-kik-kik*. **Habitat** Much as Collared, but favours steppe. **Note** Passage hatched, but rare; vagrant Bahrain, Iran, Oman, Qatar, UAE, Yemen.

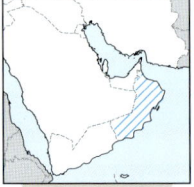

Small Pratincole *Glareola lactea* — pm, wv
L: 17. W: 45. Small, *swallow-like pratincole*. In flight, appears sandy-grey, black and white with diagnostic wing pattern: striking, *broad white secondaries with narrow black trailing edge and black primaries; underwing black with striking, broad white panel on secondaries and inner primaries*; slightly forked black-and-white tail. Flight is swift, bouncy and rather swallow-like, on long and pointed wings. Juvenile resembles adult. **Voice** Harsh notes recall Black Tern, also a Little Tern-like *tuck-tuck-tuck* or *ke-terrick-ke-terrick*; birds in feeding flocks have high-pitched rolling *prrip*. **Habitat** Dry sandy or muddy areas adjacent to wetlands, over which it hawks for insects. **Note** Passage hatched, but rare; vagrant Bahrain, Iran, UAE, Yemen. [Alt: Little Pratincole]

PLATE 50: NODDIES AND SKIMMERS

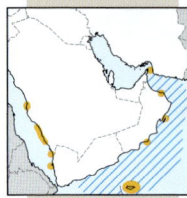

Brown Noddy *Anous stolidus*   sv, mb
L: 43. W: 83. Size of Sandwich Tern with *long, wedge-shaped tail (with shallow fork when spread) and long black bill. Dull chocolate-brown except for pale ash-grey crown, which grades into white forehead and contrasts sharply with black lores*; may appear white-capped in abraded birds in strong light; *underwing-coverts contrast clearly with blackish-brown flight feathers* (smaller Lesser Noddy has all-brown underwing). Juvenile has crown varying from grey-brown to whitish; immatures (up to three years old) are extremely abraded, appearing paler than adults. Separated from young Sooty Tern by darker underwing-coverts (can be whitish in Sooty) and shape of tail. Usually flies low over water with slower more languid wingbeats than Lesser Noddy. Feeds by hovering or banking before swooping low to snatch prey from surface; often in mixed flocks of feeding terns and Persian Shearwaters. **Voice** At breeding site deep, guttural, corvid-like *kwok-kwok, karruk* or *krao*. **Habitat** Maritime; nests colonially on rocky islets and cliffs. **Note** Present in hatched area in winter. [Alt: Common Noddy]

Lesser Noddy *Anous tenuirostris*   sv
L: 32. W: 60. Very similar to Brown Noddy *but smaller* (size of Common Tern), shorter-tailed and *with proportionately longer, thinner bill*; lacks narrow black forehead band over bill and usually has pale ash-grey (not white) forehead and crown, *lacking sharp demarcation with black lores; dark underwing without contrast between flight feathers and wing-coverts* of Brown Noddy (but reflecting light may make underwing-coverts appear paler). Juvenile generally less pale-crowned than Brown Noddy. In flight, wings narrower, wingbeats faster, the 'jizz' is lighter than the bulkier Brown Noddy, but separation difficult at distance. **Habitat** Maritime. **Note** Non-breeding summer visitor in small numbers to Masirah Island area, Oman (hatched on map); vagrant UAE, Yemen.

African Skimmer *Rynchops flavirostris*   V
L: 38. W: 106. Resembles large, dark-backed tern. *Very large orange-red bill slightly decurved* with *shorter upper mandible* and *very long, narrow, pointed wings*. Adult in summer is *blackish-brown above, including hindneck*, but forehead and trailing edge to arm white; underside white, primaries darker. Sides of short, slightly forked black tail are greyish-white. In winter and in juvenile, has a diffuse whitish collar around hindneck, making separation from Indian Skimmer difficult (see Indian Skimmer for identification). Young birds have shorter bill, tipped dark. Flight graceful with *deliberate, slow accent on upstroke*; often in flocks, following the 'leader' in a line low over the water. Skims low over water with long lower mandible breaking water surface. **Habitat** Rivers and other wetlands, rarely coastal. **Note** Vagrant Oman, Yemen.

Indian Skimmer *Rynchops albicollis*   V
L: 42. W: 108. Closely resembles African Skimmer but larger and at all ages has a *clear-cut white collar on hindneck*. African Skimmer has a less clear-cut whitish collar in winter adult and juvenile; Indian Skimmer then separated with difficulty by *all-white tail feathers with narrower dark centres*; both webs are white in all ages, whilst in African Skimmer the same feathers are white on outer webs, and greyish on inner webs; the latter become progressively paler on the outer feathers. **Habitat** Rivers and other wetlands, rarely coastal. **Note** Vagrant Iran, Oman.

PLATE 51: GULLS I

Brown-headed Gull *Choroicocephalus brunnicephalus*   V
L: 43. W: 100. Slightly larger and heavier than Black-headed Gull, with stronger bill and pale eyes. Told by *broad black wing-tip breaking white leading edge above, dark grey leading primaries below* and *small white mirror near wing-tip of both surfaces*. In summer, hood paler brown than Black-headed with clearer black rim at rear; in winter, head pattern much like Black-headed; but *iris yellow*. First-winter birds have similar wing pattern, but *broader black wing-tip without white mirror; also bold blackish trailing edge to wings*. Iris dark in juvenile but becomes pale during first winter or spring, when most of inner wing and tail moulted to adult pattern. **Voice** Deep *grarhh*. **Habitat** Coastal, lakes. **Note** Vagrant Iran, Oman, UAE.

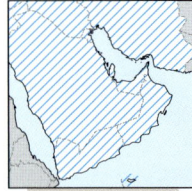

Black-headed Gull *Choroicocephalus ridibundus*   WV
L: 38. W: 93. Medium-small gull, told (except from Slender-billed Gull) by *broad white leading edge of primaries, contrasting below with dark grey remaining primaries*. In summer, has *dark brown hood* (looking blackish except close to); in winter, head white with black spot on ear-coverts and, often, vague bar from eye over crown and from ear-coverts over nape. Juvenile and first-winter have adult-type primary pattern; *told from similar Slender-billed Gull by slightly smaller size, shorter bill with darker tip, shorter neck, less bulky breast, more distinct markings on head and always dark eye*. Flight light and buoyant. **Voice** Harsh *kreeeea*. **Habitat** Coastal and inland waters; nests colonially. **Note** Passage and winter hatched; a few summer in S Arabia.

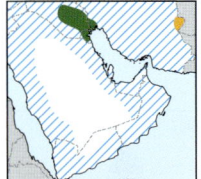

Slender-billed Gull *Choroicocephalus genei*   PM, WV
L: 43. W: 100. Larger than Black-headed Gull, *bill distinctly longer, forehead rather sloping* with long feathering at base of upper mandible accentuating length of bill. Wing pattern similar to Black-headed, but *head completely white in summer, when breast often has rosy tinge*. Legs longer; when alert *appears curiously long-necked*. In winter, often shows *small greyish spot on ear-coverts* (larger, blacker in Black-headed), *but no vague bar over crown or nape* (as shown by many Black-headed Gulls); *iris pale in winter* but dark in many breeding birds (always dark in Black-headed). *Dark red bill often appears blackish at distance*. Juvenile and first-winter also told from Black-headed by pale bill with poorly marked or unmarked tip (tip always dark in Black-headed) and different head pattern (same as winter adult); pale iris starts to show during first winter. *In flight rather hunch-backed, with more protruding neck bulging downwards* – useful characters with experience. **Voice** Hoarse nasal *yaarr*, deeper and harsher than Black-headed, can recall squeal of distant Water Rail; also abrupt cheery *yap* or *yirp*. **Habitat** Coastal and inland waters; nests colonially. **Note** Mainly summer visitor to breeding sites; passage and winter hatched.

Sabine's Gull *Xema sabini*   V
L: 34. W: 89. Smaller than Black-headed Gull with shallowly forked tail and *diagnostic tricoloured upperwing pattern*. In summer, adult has *slate-grey hood*, in winter just a blackish nape-patch; *short black bill with yellow tip*. Juvenile separated from similar Black-legged Kittiwake *by continuous brownish from crown and nape to mantle* (also breast-sides), *brownish forewing*, dusky bar on inner underwing (visible at close range), *pale fleshy-grey legs* and more forked tail (young Black-legged Kittiwake has black band at base of white hindneck, black band on inner wing, white underwing and black legs). Flight light; often swoops to water surface for food. **Habitat** Maritime, occasionally driven onshore. **Note** Vagrant Iran, Oman, UAE.

Black-legged Kittiwake *Rissa tridactyla*   V
L: 40. W: 95. Slightly larger than Black-headed Gull. Adult resembles Common Gull but has *all-black wing-tip without white spots*, darker grey back and inner wing but *whitish, translucent outer wing and black legs*. First-winter recalls much smaller Little Gull, but head white (just small black spot behind eye) with *bold, black band on hindneck, white secondaries and slightly forked tail*. First-summer birds often lack band on hindneck and sometimes lack bar on inner wing. **Habitat** Coastal and offshore waters. **Note** Vagrant Iran, Oman, UAE.

PLATE 52: GULLS II

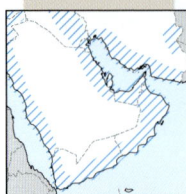

Pallas's Gull *Ichthyaetus ichthyaetus* WV
L: 68. W: 158. Very large; adult in summer unmistakable. In winter, has large dusky patch behind eye; *white eye-crescents present in all ages. At rest appears deep-chested, with long sloping forehead accentuating length of bill.* Readily identified in flight, even at extreme range, by extensive pale-based primaries (upperwing thus tricoloured). Juvenile and first-winter from other young large gulls by *unmarked white rump and tail with clear-cut band at tip*, unmarked white underparts (though juvenile has grey-mottled breast-band or patches at sides, lost in first winter), white underwing with extensive black wing-tips and often black-tipped coverts forming underwing lines, *pale mid-wing panel above, and head pattern*. First-winter told from second-winter Caspian Gull by size, *sharp tail-band*, head shape and pattern, dark-mottled hindneck, darker inner primaries and longer bill. Second-winter still shows fairly distinctive tail-band. **Voice** Loud and deep but hoarse, strangulated *kra-ah*. **Habitat** Coastal, mostly rare inland. **Note** Passage and winter hatched. [Alt: Great Black-headed Gull]

White-eyed Gull *Ichthyaetus leucophthalmus* pm
L: 40. W: 108. From heavier Sooty Gull by *long and slender all-dark bill* (dark red with black tip in adult; blackish in immature), which droops at the tip (bill stouter and bicoloured in Sooty); *hood and bib black* (dark brown in Sooty), *upperparts greyer, with conspicuous white eye-ring at all ages* (faint and narrow in Sooty). Immature similar to Sooty Gull though upperpart feathering less conspicuously pale-fringed. At distance can be mistaken for a skua. **Voice** As Sooty Gull but less harsh and deep. **Habitat** Coastal; nests colonially on low-lying islands. **Note** Dispersal throughout hatched area; rare Oman; vagrant Iran, UAE.

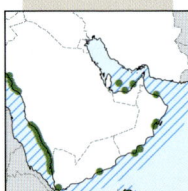

Sooty Gull *Ichthyaetus hemprichii* RB, WV
L: 44. W: 112. Told from similar White-eyed Gull by *dark brown hood and bib* and *thick, straight bill, which is bicoloured (at all ages)* – in adult yellow-green with black band behind red tip; in immature pale blue-grey with black tip. *Narrow white crescent above eye* at all ages (seldom also below). Immature brownish-grey on head, breast and flanks; upperparts brownish with buff feather-edges. Rather long-winged, relatively short-tailed and front-heavy but with steady flight; underwing dark. At distance over sea can be mistaken for skua; often piratical on other gulls and terns. **Voice** Loud mewing *kaarr* or *keee-aaar*; also high-pitched *kee-kee-kee*. **Habitat** Coastal; often near ports and fishing villages; nests on islands or cliffs, mostly colonial. **Note** Dispersal throughout hatched area; vagrant Bahrain.

Common Gull *Larus canus* pm, wv
L: 43. W: 115. Recalls *small Caspian Gull, but wings narrower and bill distinctly smaller, thinner and greenish-yellow without red spot*. Black wing-tip with prominent white spots, larger area than in Caspian. In winter, bill has narrow black band near tip; eyes dark. Juvenile resembles Mediterranean Gull, but with greenish bill, mottled breast, brown markings on underwing and less contrasting upperwing pattern. *In first-winter, back blue-grey*, upperwing pattern less contrasting and bill bicoloured. **Voice** Alarm call *klee-uu* with emphasis on last high-pitched note. **Habitat** Coastal and inland waters; sometimes fields. **Note** Passage and winter hatched, rare in south; vagrant Bahrain, UAE. [Alt: Mew Gull]

PLATE 53: LARGE WHITE-HEADED GULLS

Lesser Black-backed Gull *Larus fuscus*   WV, PM

Several subspecies – *fuscus*, *heuglini* and *barabensis* – occur in the region and are a source of much confusion. Some observers simply refer to these and the Caspian Gull as 'large white-headed gulls'. However, with care, experience and close attention to wing, leg and eye colour and, particularly, to wing-tip pattern, it is possible to separate them, at least adult birds. The three subspecies are described separately below.

Baltic Gull *Larus fuscus fuscus*

L: 53. W: 127. Smallest of the group, a slim, very long-winged gull. *Adult with jet-black wings and mantle, though can wear greyer*, rather sparse streaking on head and neck in winter; *bright yellow legs and one, rarely two, small, white mirrors* on wing-tip. In flight, wings noticeably long (also evident on settled bird) and slender. *Little or no contrast in black tone on upperside of primaries*. Flying first-winter shows *evenly coloured dark brown 'hand', secondaries and greater coverts without translucent inner primaries*, unlike Caspian; tertials dark brown, narrowly fringed and tipped buffish-white; tail and rump white, tail-base boldly spotted or barred black and with broad, blackish terminal band. *Underwing-coverts dark brown in first-winter and usually lacking distinctive pattern* (unlike Caspian which are plain white). As immature plumage progresses, head and underparts turn whitish, and from spring of second calendar year blackish feathers are visible on shoulder, back and wing-coverts. **Voice** Calls deep and slightly nasal. **Habitat** Coastal; less frequently inland waters. **Note** Passage hatched; scarce or absent in winter.

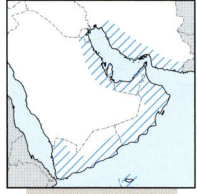

Heuglin's Gull *Larus fuscus heuglini*

L: 58–65. W: 125–150. Large, with upright stance; recalling Steppe Gull but with fiercer face; *one, rarely two white mirrors* on wing-tip. Bold streaking on head and neck in winter to early April (to February in Steppe Gull). Juvenile can appear similar to Baltic Gull. First-winter/first-summer can appear like Caspian Gull, but often has bicoloured bill, patterned greater coverts, more heavily marked scapulars, and less translucency on inner primaries. **Voice** Deep, nasal *gagaga*; usually silent in winter. **Habitat** Coast and lakes. **Note** Passage and winter hatched; abundant Oman, UAE. [Alt: Siberian Gull]

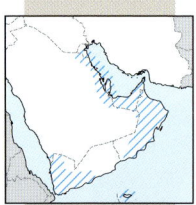

Steppe Gull *Larus fuscus barabensis*

L: 55–65. W: 125–150. Slightly smaller than Caspian Gull with more rounded head, flattish forehead, medium/long bill, shorter legs and brighter bare parts; *smallish, usually dark eye* – birds that winter in Arabian Gulf often have pale eyes. Usually darker grey than Caspian, slightly paler than Heuglin's Gull. *Grey tongues protrude into black wing-tip less than Caspian, but more than Heuglin's; mirrors on P9 and P10 (longest primaries). Underside of flight feathers grey against white coverts* (unlike most Caspian but similar to Heuglin's); legs generally deeper yellow in breeding condition, but some fleshy. May show band on bill. Winter adult white-headed with weak streaks on hindneck. First-summer has mantle, scapulars and many coverts of adult type. **Voice** Gulping *yah-aah-aah-aah* and thin, high, drawn-out *peeeer* in winter. **Habitat** Coastal in winter. **Note** Passage and winter hatched; abundant Oman, UAE.

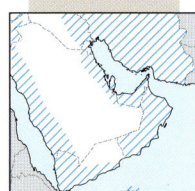

Caspian Gull *Larus cachinnans*   pm, wv

L: 60–65. W: 125–150. Long-winged, long-legged, with *long, parallel-sided bill and upright stance. Neat looking with characteristic blank or innocent facial expression, gently sloping forehead, rounded head and often a dark, bullet-hole eye*. Adult paler than all likely congeners (except Pallas's Gull). *More white on wing-tip than other gulls*, with large white tongue on inner web of P10 (outermost primary) and often with large all-white tip; some eastern populations have small amount of black within the large white wing-tip. Grey tongues protrude into more restricted black on primaries. Legs pale greyish-pink/straw; bill with a greenish tone to the base. Juvenile quickly moults to first-winter, a beautiful gull with a clean white head, nestled in 'shawl' around hindneck, and white underparts; pale tips to median and greater coverts form pale borders to finely patterned greater coverts; underwing normally white. Subtle pale line from eye to eye, along nape/crown boundary, with 'shadow' behind eye. First-winter Heuglin's Gull can look similar to Caspian Gull. **Voice** Nasal and ringing, likened to a donkey braying. **Habitat** Coastal and inland waters. **Note** Passage and winter hatched; least numerous of the larger gulls in much of Arabia.

PLATE 54: LARGE WHITE-HEADED GULLS – COMPARISON PLATE

The plate opposite shows the plumage for a typical adult of the three most confusing 'large white-headed gulls' (Heuglin's, Steppe and Caspian).
The table below provides comparisons of the adult plumage characteristics, bare part coloration and moult timings of the 'large white-headed gulls', as well as giving additional tips on identification. Care is needed in assessing the upperpart grey tone of adults, as shade changes with light and angle.

Species	Status	Back Kodak greyscale	Leg colour*	Bill**	Wing-tip (best seen in flight)	Helpful tips
Heuglin's	Abundant, WV, PM	8–11 (darkest grey)	Yellow	Yellow	Some contrast with rest of wing; small mirror P10 and usually P9	Fierce face; may keep head-streaks till early April, cf. Steppe
Steppe	Abundant, WV, PM	7–8.5	Yellow	Often warm, bright yellow	Between Heuglin's and Caspian, smaller grey tongues than Caspian, subterminal mirror P10, smaller P9; smaller tongue on underside of P10	Kindly face; often dark, bullet-hole eye; loses head-streaks by mid-February, cf. Heuglin's
Caspian	Uncommon, pm, wv	4.5–6.5 (palest grey)	Pale greyish-pink/straw	Long, parallel-sided, often greenish at base	Upper: narrow grey tongues into black; mirror P10 merges with white tip, smaller mirror P9. Under: large white tongue on inner web of P10	Kindly face, long spindly legs, parallel-sided wings

Note that in all these gulls males are larger than females with larger bills and a more fierce expression.
* Gulls with yellow legs can, in winter or out of breeding condition, have pinkish legs.
** Adults of all species have a yellow bill with a red gonys spot, and all can show black near the bill-tip.

Species	Mar	Apr	May	June	July	Aug	Sep	Oct	Nov	Dec	Jan	Feb	Mar
Heuglin's						arrests moult							
Steppe					arrests moult								
Caspian													

starts primary moult   continues primary moult

Adult primary moult progression interpreted from Olsen K. M. & Larsson H. (2004) *Gulls of Europe, Asia and North America*, Christopher Helm, London.
Note: Northern breeding species moult later than southern species.

PLATE 55: LARGER TERNS

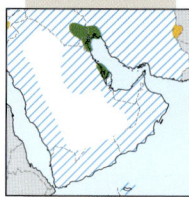

Gull-billed Tern *Gelochelidon nilotica* — PM, WV
L: 38. W: 95. Medium-sized tern somewhat resembling Sandwich Tern *but lacks crest and has shorter, thicker, gull-like all-black bill, shallower forked tail and whitish-grey rump and tail* (white in Sandwich); lacks contrast shown by Sandwich Tern between dark outer and pale inner primaries above; outer primaries have distinct dark trailing edge. In winter, lacks black cap *but has variable black 'mask', behind eye*. Juvenile is much less spotted on scapulars and innerwing-coverts than similar Sandwich, and has white crown and *black eye-patch, recalling winter adult Mediterranean Gull*; in first-winter head even whiter. Flight more leisurely and aerobatic than Sandwich. When standing, note more robust build with *long black legs*. Feeds over fields, marshes, dipping to surface. **Voice** Nasal *ger-wek* with stress on second syllable; alarm fast, agitated, nasal laughing notes *dididit*; juvenile call soft *pre-eep*. **Habitat** Sheltered coasts, inland waters; nests colonially. **Note** Passage hatched; winters mainly Iran and S Arabia.

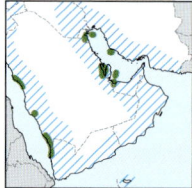

Caspian Tern *Hydroprogne caspia* — PM, WV
L: 53. W: 135. Near size of a large gull, identified by *large red bill, almost gull-like flight with slow steady wingbeats and by call; distinctive dark primaries below*. Juvenile has orangey bill, more extensive black cap to below eye (unlike adult), weak dark scales on mantle and wing-coverts, primaries dark both above and below; legs pale (black in adult). **Voice** Deep, loud, harsh and deliberate *aark*, recalling Grey Heron, or longer retching *kraa-jak*. **Habitat** Open coasts, lagoons; inland wetlands; nests in winter in Arabian Gulf, singly or colonially on small islands. **Note** Passage hatched; winters mainly in Arabia.

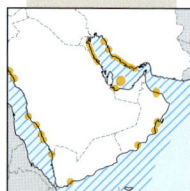

Greater Crested Tern *Thalasseus bergii* — RB, PM, WV
L: 46. W: 105. Large, between Sandwich and Caspian Terns in size. Adult from smaller Lesser Crested Tern *by longer, thicker, more drooping, waxy greenish-yellow bill and by darker grey upperpart*s, with rump and tail paler grey than back; tail slightly longer, more forked, and wings broader than Lesser Crested. In summer, black crested cap always broken by white forehead-band (never black as in some Lesser Crested). In winter, bill paler greenish-yellow, or washed-out yellowish-orange; crown white with black confined to nape; tail and rump almost as dark as back. Juvenile has pale innerwing-panel above, framed by dark bar on forewing and dark covert- and secondary bars. Flies with shallow, deliberate wingbeats. **Voice** Deep, rough, *kee-rit*, or gravelly *craark*; also a high-pitched *kree-kree*. **Habitat** Coastal; nests colonially on sandy or rocky islands. **Note** Disperses to hatched areas. [Alt: Swift Tern]

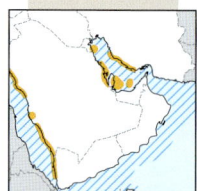

Lesser Crested Tern *Thalasseus bengalensis* — PM, WV
L: 41. W: 92. Size as Sandwich Tern. *Long, slim orange-yellow bill, upperparts pale ash-grey, with rump and tail paler grey*. Crested nape black; forehead in summer is white or black. In winter bill paler orange-yellow; solid black confined to nape. Juvenile has bill paler orange than adult; resembles juvenile Swift Tern but upperwing has paler dark bands and less pronounced pale mid-wing panel; dark markings on inner wing retained to first spring as are dark primaries above. *Told from Sandwich Tern by bill colour* and slightly darker upperparts. **Voice** Recalls Sandwich Tern closely, but less grating and not so obviously disyllabic, *krriik-krriik* or *kreet-kreet*; also *kir-eep* and *kee-kee-kee*; juvenile call much as juvenile Sandwich Tern. **Habitat** Coastal; nests colonially on sandy or rocky shores or islands. **Note** Disperses to hatched areas.

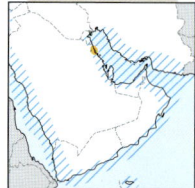

Sandwich Tern *Thalasseus sandvicensis* — PM, WV
L: 41. W: 92. Medium-sized with *pale grey upperparts and wings (which may look almost white at sea), long slender black bill tipped yellow* (yellow tip hard to see at distance), long narrow wings, medium-short forked tail and fairly long neck in flight, which is powerful with deep wingbeats. In summer black cap has ragged crest; in winter, black confined to nape extending forward to eye (like Lesser Crested Tern). Juvenile has all-black bill, dark-scaled mantle and forewing-coverts, but lacks the broad dark bars on inner wing of similar-aged Lesser Crested; first-winter has grey mantle and scapulars and head pattern like winter adult. **Voice** Loud, grating disyllabic *kerr-rick*; juvenile's call heard into winter, high-pitched *k-rill*. **Habitat** Coastal. **Note** Passage and winter hatched; many non-breeders oversummer.

PLATE 56: SMALLER TERNS

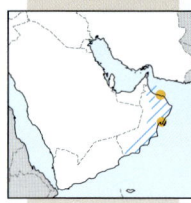

Roseate Tern *Sterna dougallii* — mb, sv

L: 38 (including tail streamers). W: 77. Similar to Common Tern but *adult paler above; whiter below*, tinged pink in breeding season. *Long, thin bill largely blackish* in spring but becomes red by July; *tail streamers very long and white*. Wing pattern diagnostic at all ages — *dusky outer primaries lacking black line along trailing edge of outer primaries below*. At rest, tail streamers protrude far beyond tip of tail; legs longer than Common. Winter plumage similar to Common but retains long tail streamers. Juvenile, which has short tail streamers, told from Common/Arctic by underwing pattern (white trailing edge to outer primaries below); also has darker forehead than Common/Arctic and more scaly upperparts and scapulars. Flight fast and direct with stiffer, more rapid wingbeats than Common. **Voice** Characteristic, soft guttural *cherr-wrick*. **Habitat** Coastal. **Note** Passage hatched; vagrant Bahrain, Saudi Arabia (Gulf), UAE.

Common Tern *Sterna hirundo* — PM, wv, sv

L: 35. W: 80. Closely resembles Arctic Tern, with wings slightly broader, tail streamers shorter, *bill and legs longer*, forehead flatter and wingbeats often more powerful. *Dark outer primaries above abruptly cut off from grey inner primaries* (uniformly pale grey in Arctic); *from below has blackish band on trailing edge of outer primaries and translucency confined to innermost primaries* (all translucent in Arctic); *bill dark orange-crimson, usually tipped black*; cheeks and throat whiter than in Arctic. Bill of adult in subspecies *minussensis* appears *all black in summer* (close to up to half of basal half and cutting edges deep red); also has *white chin and throat but light grey breast and belly*, though grey paler than in breeding White-cheeked Tern, from which it also differs in larger size, characteristic *white rump, and differing upper- and underwing pattern*. Juvenile best told from similar Arctic by flight, *dark grey secondaries* above (white in Arctic), broader, darker forewing-band above, and translucency and pattern of primaries (similar to adult); centre of rump pale grey (clean white in Arctic). **Voice** Deeper than Arctic; sharp *kitt*; *kirri-kirri-kirri*, and a drawn-out *kreee-aeh*. **Habitat** Coastal and inland waters. **Note** Passage hatched; winters mainly in S Arabia. Subspecies *minussensis* is regular on passage in Arabian Sea and The Gulf (subspecies not separable in winter plumage).

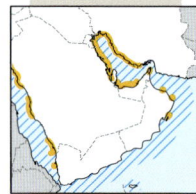

White-cheeked Tern *Sterna repressa* — MB, PM

L: 33. Slightly smaller than Common Tern (often mixes on passage/winter) with shorter wings, slightly shorter legs, and *more slender bill, proportionately shorter than in Common Tern* and usually slightly downcurved. Adult in summer is *dark silver-grey above with grey underbody; underwing with whitish area in centre (also evident in juveniles)* and *primaries above appearing paler than inner wing (both visible, and diagnostic, at considerable range); white cheek-stripe* recalls adult Whiskered Tern in summer (but tail-fork and bill longer); secondaries dull grey, with long, broad black line to tips of outer primaries below. In winter, remains *dull grey above (slightly darker than Common), including rump and tail*; underparts white, mottled with dark grey in some; bill blackish. First-winter has broad blackish forewing-band and dark secondaries (like Common), but mid-wing dull greyish rather than greyish-white as in Common; underwing similar to adult, with rump and tail greyish (pale grey on centre of rump only in Common). Settled juvenile/first-winter birds resemble marsh terns. **Voice** Often gentler than Common; loud *kee-err* or *ker-rit* with emphasis on short second syllable (on first note in Common) and single *kip*, less sharp than Common. **Habitat** Coastal and maritime; nests colonially on bare islands. **Note** Passage hatched; scarce in S Arabia in winter.

Arctic Tern *Sterna paradisaea* — V

L.: 38. W: 80. Similar to Common Tern but rounder head, shorter thicker neck, longer tail streamers and more elegant, buoyant flight. *Bill dark red, usually without black tip; legs shorter. In flight all primaries appear translucent, revealing a thin, sharp black line to trailing edge of outer primaries*. In summer, greyish underparts with whitish band below black cap. Juvenile/first-winter has *white secondaries*, diffuse dark bar on leading forewing and white rump. **Voice** Similar to Common Tern, but harsher. **Habitat** Coastal; occasionally inland. **Note** Vagrant Kuwait, Oman, UAE.

PLATE 57: SMALL AND PELAGIC TERNS

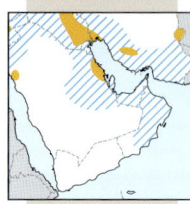

Little Tern *Sternula albifrons* mb
L: 23. W: 53. Small with *fast wingbeats*. Adult in summer is pale grey above, extending in some onto centre of otherwise *white rump*; *usually two grey-black leading primaries (with white shafts)*; *white forehead usually extending in point to rear of eye*; *legs bright yellow to reddish-orange*. In winter, bill black, legs dull grey or brown, usually with some yellow; black on head reduced to band around nape; rump and tail largely grey. Juvenile (dark bill with reddish base) has dark 'U'-shaped markings on pale grey mantle, scapulars and tertials, lost in first winter when resembles winter adult; legs greyish-pink to yellow-brown; upperwing shows dark outer primaries becoming progressively paler inwards; leading forewing dark. **Voice** Excited, hoarse grating *kryik* or *pret-pret*. **Habitat** Coastal and inland waters; nests on beaches or sandbanks in rivers. **Note** Passage hatched; winter range obscured by confusion with Saunders's Tern; vagrant Yemen.

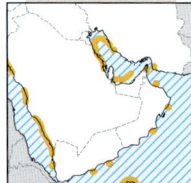

Saunders's Tern *Sternula saundersi* RB, PM, WV
L: 22. Separation from Little Tern problematic. In summer, upperparts contrast markedly with *blacker outer three primaries (dark-shafted)*; *white forehead not usually reaching eye* thus squarer; *rump more extensively grey than in Little and concolorous with back*; *legs generally darker, reddish- or pinkish-brown*, sometimes some yellow on rear tarsus. In winter, adult and first-winter birds doubtfully separable (Saunders's has *darker grey upperparts* than similar-aged Little in Red Sea; apparently not so in Arabian Gulf). **Voice** Strident calls recall Little closely, but often lacking same urgency; sharp *kip* or *wip*, and excited *tchijjick*. **Habitat** Coastal, rarely inland; nests on beaches. **Note** Winters on Iranian and Arabian coasts.

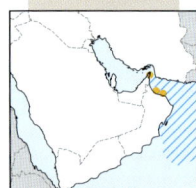

Bridled Tern *Onychoprion anaethetus* MB, PM
L: 37. W: 76. Medium-sized slender tern, Sooty Tern being the confusion species. *Dark ashy upperparts* (but often appearing blackish), *narrow white forehead-band extending behind eye as narrow supercilium*, black cap contrasting with *pale grey hindneck-collar* and greyer mantle (often hard to see in flight); long outer feathers of deeply forked tail have white sides. Whitish underparts and underwing-coverts contrast with *dark silver-grey central tail feathers and especially flight feathers*. Underparts can appear grey. Juvenile lacks distinct supercilium, has pale grey-brown back and wing-coverts, edged buffish. Flight graceful with slimmer wings than Sooty Tern. Does not dive but takes food from surface of water. May settle on sea, with tail pointing upwards at 45°. **Voice** High-pitched *kee-yharr*, yelping *wep-wep* and a harsh grating *karr*; also a pleasing rolled *purrurr* or *prerrr*, given day or night at sea or in colony. **Habitat** Maritime; breeds colonially on rocky or sandy islands, in crevices or under low vegetation. **Note** Passage hatched; winters in Arabian Sea.

Sooty Tern *Onychoprion fuscata* mb
L: 44. W: 90. Larger, more stocky than Bridled Tern and with *entire upper surface uniform blackish* (paler mantle and hindneck in Bridled), *broader but shorter white forehead-patch than Bridled (not extending beyond eye)*, and black loral streak reaching gape (not base of upper mandible as in Bridled); white underparts and underwing-coverts *contrast more with dark grey flight feathers, outer primaries lacking whitish wedge towards wing-tip*. Juvenile *sooty-brown all over* except for whitish belly and undertail-coverts; upperparts flecked whitish. Told from young Brown and Lesser Noddies by whitish vent and undertail-coverts, paler underwing and forked tail. First-summer like adult but throat and upper breast blackish. May settle on sea surface, as Bridled Tern. **Voice** Call diagnostic, a high-pitched *ker-wacki-wah*. **Habitat** Maritime; nests colonially in open on sandy, rocky or vegetated islands. **Note** Dispersal as hatched, but rare; vagrant Bahrain, Iran, UAE.

PLATE 58: MARSH TERNS

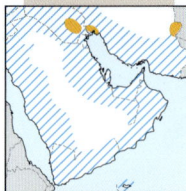

Whiskered Tern *Chlidonias hybrida*   PM, WV, sv
L: 25. W: 73. Slightly larger than White-winged Tern with broader wings, more forked tail (adult), longer legs and *heavier bill*. Flight buoyant but more *Sterna*-like; dips or dives for food. Adult in summer has *sooty-grey underparts contrasting with whitish cheek-stripe and underwing*; in winter separated from White-winged *by flight action, larger bill, streaked crown and shape of ear-coverts patch*. Juvenile from similar White-winged by head pattern, practically no white collar on hindneck, paler 'saddle' and pronounced blackish and buffish markings on scapulars (contrasting with 'saddle'; scapulars dark in White-winged); upperwing slightly paler with virtually no dark bar on leading forewing; rump and tail usually concolorous pale grey (rump white in White-winged); sometimes has dark smudge at sides of breast, as some adults in winter, absent in White-winged. **Voice** Loud, hoarse *kreck*, almost a sneeze. **Habitat** Inland and coastal waters. **Note** Passage and winter hatched; rare in north in winter.

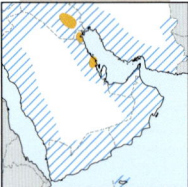

White-winged Tern *Chlidonias leucopterus*   PM, WV, sv
L: 22. W: 65. Summer adult distinctive; in winter from similar Black Tern *by lack of dark patch at sides of breast*, whiter rump and tail (pale greyish in Black); a few show some black marks on underwing-coverts, making identification easier. *Juvenile has more contrasting darker 'saddle' and paler upperwing than juvenile Black, and lacks dark breast-patch* and white tips to rear scapulars. Rump and sides of tail whiter, with dark patch on ear-coverts extending further downwards. Moults dark 'saddle' to pale grey in late autumn. **Voice** Harsh, dry *kesch* or *kruek*, less hoarse than Whiskered. **Habitat** Inland and coastal waters. **Note** Passage hatched; scarce in winter when mainly in S Arabia. [Alt: White-winged Black Tern]

Black Tern *Chlidonias niger*   pm, sv
L: 23. W: 66. Summer adult distinctive; in winter underparts white with *dark patch on sides of breast*, more pronounced in juvenile; forehead and hindneck-collar white; upperparts greyish, darker grey and more uniform than White-winged. Juvenile has dark fore-mantle and leading upperwing-coverts; scapulars tipped whitish; sides of rump sometimes whitish, rest of rump and tail pale greyish. First-winter recalls adult in winter. **Voice** Short sharp *kjeh* and repeated *kit*, more squeaked than sneezed as in congeners. **Habitat** Inland and coastal waters. **Note** Passage hatched; vagrant Oman and in The Gulf.

PLATE 59: SKUAS

South Polar Skua *Stercorarius maccormicki* — V

L: 53. W: 127. Closely resembles Brown Skua. Pale and intermediate morphs identifiable by *creamy buff or grey-brown head, hindneck and underparts, contrasting with blackish underwing-coverts*. Dark morph often has pale forehead, hindneck with pale wash at sides and bill sometimes more bicoloured. During primary moult (July–August) immature may have *conspicuous symmetrical gaps in each hand*. Pale 'noseband' at base of upper mandible shown by some birds. **Habitat** Maritime. **Note** Vagrant Oman, Yemen.

Brown Skua *Stercorarius antarcticus* — V

L: 63. *Larger and darker than South Polar Skua*, with broad-based wings, *heavier bill*, and lacking contrasting paler head and underbody, both species showing conspicuous white primary patches (above and below) even at long range. Adult very similar to dark morph of South Polar Skua, but upperparts less uniform with *prominent pale mottling and flecking* (but only visible at close range). Juvenile warmer in tone without heavy mottling above. Some individuals cannot be safely told from South Polar Skua. Typically thicker-billed and darker chocolate-brown than extralimital Great Skua *Stercorarius skua*, but latter not a reliable or constant field character. **Habitat** Maritime. **Note** Vagrant Oman; vagrant unidentified large skuas in Iran and UAE also considered to be this species.

Pomarine Skua *Stercorarius pomarinus* — PM, WV, SV

L: 51–56. W: 125. Size of large gull, *heavily built* and, in adult, *elongated broad and twisted central tail feathers* ('spoons') diagnostic, though these are often lost. Two morphs: dark (scarce) and pale morph, which differs from Arctic Skua in more extensive black cap, darker flanks and blacker breast-band (sometimes absent) and vent. In winter, flanks and tail-coverts barred, elongated tail feathers often short and always blunt-ended. Juvenile best separated from similar Arctic Skua by *more regular bars on flanks, vent, rump and tail-coverts*. Whitish base to greater primary coverts below often conspicuous, even at range (frequently absent in Arctic Skua). Bill proportionately larger, heavier at base and more distinctly bicoloured. With experience can be *identified by larger size, slower wingbeats, broader wings, deeper breast and belly*; also, in juvenile, when present, short, blunt central tail feathers (pointed in Arctic Skua). Piratical attacks more direct, less agile than Arctic, frequently chasing large gulls. **Habitat** Maritime. **Note** Passage hatched; winters mostly in Arabian Sea. [Alt: Pomarine Jaeger]

Arctic Skua *Stercorarius parasiticus* — PM, WV, SV

L: 46. W: 117. Streamlined appearance in flight, appearing intermediate in size and proportions between Pomarine and Long-tailed Skuas. Adult in summer has *elongated, pointed central tail feathers* (6–11cm). Two morphs occur (dark and pale), with frequent intermediates. Juvenile variable, from pale to very dark birds; ground colour generally warmer, more rusty than young Pomarine Skua; juvenile told with experience from young Long-tailed Skua by *broader wings, slightly shorter tail* (equals width of wing-base), *shorter, pointed central tail feathers* (longer, blunt-ended in Long-tailed), *thicker rear-body, less black on bill-tip and warmer brownish ground colour* (colder, greyer in Long-tailed Skua), *with rusty fringes and barring*. Normal flight steady, falcon-like, straight and fast; piratical attacks with sudden twists and turns, harassing birds the size of Sandwich Tern or Common Gull. **Habitat** Maritime. **Note** Passage hatched; winters mostly in Arabian Sea. [Alt: Parasitic Jaeger]

Long-tailed Skua *Stercorarius longicaudus* — V

L: 53. W: 111. Slim and lightly built; size of Black-headed Gull. Resembles pale and rather grey Arctic Skua but adult told by *very long, flexible central tail feathers* (16–24cm), *lack of breast-band, white forebody gradually darkening towards rear, pale greyish upperparts contrasting with blackish flight feathers with pale shaft streaks only on outermost primaries*, and *pale blue-grey legs*. Juvenile variable like young Arctic Skua, but has *colder, greyer ground colour and fringing* (never rusty), *more distinct barring on uppertail-coverts* (except in darkest birds), *thin white shaft-streak in leading primaries* (hardly visible beyond 500m, unlike distinct white flash on most Arctic Skuas) and *blunt-ended central tail feathers*. Primaries on settled bird are plain (broad buffy tips on Arctic). At rest shows more rounded head and shorter, proportionately thicker *bill with black tip to half the length*, unlike Arctic Skua. Also relatively deeper-chested with thinner rear-body. Flight more buoyant and tern-like than Arctic Skua; includes more circling and hovering, without making aggressive piratical attacks. **Habitat** Maritime. **Note** Vagrant Iran, Kuwait, Oman, Qatar, UAE. [Alt: Long-tailed Jaeger]

PLATE 60: SANDGROUSE

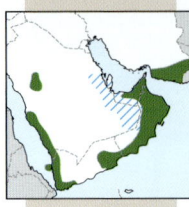

Chestnut-bellied Sandgrouse *Pterocles exustus*   RB, WV
L: 32. W: 50. In fast flight shows *diagnostic all-dark underwing continuous with dark chestnut belly*. Female shows black-spotted breast and narrowly vermiculated blackish upperparts; male has golden wings with some dark barring; both sexes show *elongated tail and narrow black bar on lower breast*. Juvenile smaller, with reduced area of dark on belly, which can often be difficult to see. **Voice** Far-carrying, short, rather guttural but liquid *kwit-kwit-kwituroh-kwituroh-kwituroh* or *gattar-gattar*; also *ke-rep, kerep* with stress on last syllable. **Habitat** Sand and gravel semi-deserts, though often near agricultural land; also coastal dunes, inshore islands. Visits water to drink in large flocks in early to mid-morning. **Note** Nomadic; winter dispersal hatched; vagrant Kuwait.

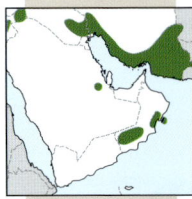

Spotted Sandgrouse *Pterocles senegallus*   RB
L: 32. W: 59. A pale sandgrouse *with a long pointed tail*; in flight shows rather *pale upperwing including primaries* with indistinct dark rear border; underwing pale with contrasting dark flight feathers (especially secondaries) which separates from Chestnut-bellied Sandgrouse. Bold spots on upperparts and breast in female, confined to wing-coverts and shoulder in male. Flocks, often large, visiting water-holes in morning; individuals occasionally join Chestnut-bellied flocks. **Voice** In flight, a disyllabic bubbling whistle, frequently repeated: *wi-dow*, first syllable higher. **Habitat** Mainly sandy deserts; also semi-deserts with sparse vegetation or scrub. **Note** Mainly resident. Vagrant UAE.

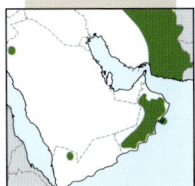

Crowned Sandgrouse *Pterocles coronatus*   RB
L: 28. W: 57. Rather pale with *short tail*. Male has yellowish head and neck, and diagnostic *black mask on forehead and around bill base*. Female sandy, finely vermiculated; head, neck and throat unmarked yellowish. In flight, *black flight feathers contrast with sandy-grey upperparts and wing-coverts* (unlike Spotted Sandgrouse); underwing-coverts white. Distinguished from Spotted Sandgrouse by short tail, which is white-tipped (noticeable when tail spread on landing) and distinctive upperwing pattern; Lichtenstein's Sandgrouse is more barred and lacks yellowish on head. Comes to water mainly in morning. Two subspecies occur in Arabia: *atratus* in sand desert and *saturatus* in the foothills of the Hajar mountains, Oman. **Voice** Frequently calls in flight on way to or from water: hard, accelerated, nasal *kaaa-kata-kata-kataah*, rendered as 'gebäcked potato'. **Habitat** Stony and semi-deserts. **Note** Mainly resident. Historical reports, but no recent records, in UAE.

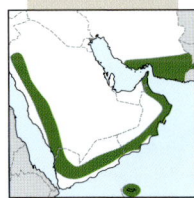

Lichtenstein's Sandgrouse *Pterocles lichtensteinii*   RB
L: 25. W: 50. Small with *short square-ended tail*. Yellowish-buff plumage *finely vermiculated black and white all over*. Male has *yellowish breast-patch framed by two black bars*; bill orange, *white forehead with vertical black bars*. Female duller, lacking distinct markings on head and breast; lacks yellow throat of similar Crowned Sandgrouse. In flight, upperwings show black flight feathers and pale wing-coverts, underwing pale with slightly darker flight feathers. Singles or small parties drink at dusk or before dawn. Active by night, when calls frequently in flight. **Voice** Call, when coming to drink, is a repeated disyllabic, melodic *whee-ak*, with stress on first syllable; when flushed a harsh, whirring *arrk*; at night, in flight, a clear liquid *whit, wheet, wheeoo*. **Habitat** Rocky deserts, *arid mountains*, wadis and hillsides with sparse scrub. **Note** Resident, but avoids Empty Quarter.

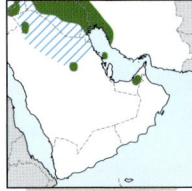

Pin-tailed Sandgrouse *Pterocles alchata*   V
L: 35. W: 60. Large, plump sandgrouse, *with long tapering tail*. In flight, *white underparts contrast sharply with black primaries and chestnut-buff breast-band (male) bordered narrowly with black*; female has pale golden-yellow breast-band with *three black bars*. Male has black throat and eye-stripe and spotted grey-green upperparts; female duller, mottled and barred above. Gregarious, often in large noisy flocks. **Voice** Call in flight a distinctive, repeated, slightly falling *arrrh, arrrh* or shorter *arrk-arrk-arrk-*. **Habitat** Dry plains and stony semi-deserts, fields. **Note** Nomadic or partial migrant as hatched; introduced UAE, vagrant Oman.

PLATE 61: PIGEONS AND TURTLE DOVES

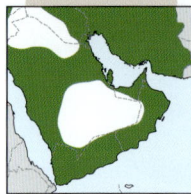

Rock Dove *Columba livia* RB
L: 33. Pale blue-grey pigeon, recalling Stock Dove, with *two broad black bands across secondaries above; rump white* or grey (rump grey in most Arabian populations); underwing white with dark band at rear; black band on outer tail more pronounced than in Stock Dove. Rock Dove is the ancestor of the familiar Feral Pigeon, which can be blackish, whitish or even reddish; most show black bands on wing, white underwing and contrasting tail pattern. Gregarious, flight very fast and often aerobatic, gliding on lifted wings. **Voice** Cooing *kru-oo-u*, second syllable stressed and highest in pitch. **Habitat** Rocky wadis, sea cliffs, mountains; nests in cave or rock ledge. Feral Pigeon widespread, often in towns; nests on buildings and cliffs. **Note** Pure Rock Dove populations still believed to exist in Oman.

Stock Dove *Columba oenas* V
L: 33. Medium-sized and easily told from Common Wood Pigeon by *absence of white band on upperwing*, also smaller, more compact and shorter-tailed, with quicker wingbeats. From Rock Dove by less white underwing *with darker flight feathers, pale ashy-grey wing-panel above and lack of bold black bars across secondaries*. Rump grey. Bill pale-tipped (dark in Rock Dove); iris dark. Often in small flocks. **Voice** Monotonous muffled, hollow 'cooing' with emphasis on first syllable, *uu-rur... uu-rur... uu-rur*; also similar single *ur... rer*. **Habitat** Open wooded areas with old trees; nests in hole in tree, rock or building. In fields on passage and winter. **Note** Winter hatched, rare in south; vagrant Bahrain, Kuwait, Oman, UAE.

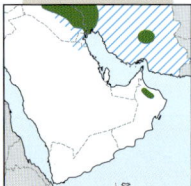

Common Wood Pigeon *Columba palumbus* rb, wv
L: 41. Large stocky pigeon, told by size, *bold white band on upperwing in flight* (all ages) and *white patches on sides of neck* (absent in young). Flight slower and heavier than other pigeons, with tail proportionately longer. Gregarious outside breeding season. Breeding birds in Oman and SW Iran, *iranica*, have white neck-patch, while *casiotis* (SE Iran) has smaller, buffy neck-patch. **Voice** Hoarse cooing, *cu-cooh-cu, coo-coo*, second note stressed and drawn out. **Habitat** Juniper woodlands above 2,000m in summer, descending lower in winter. **Note** Winter range hatched. Localised resident in mountains of N Oman (*iranica*), but also a rare winter visitor (possibly *casiotis*).

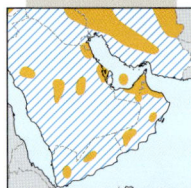

European Turtle Dove *Streptopelia turtur* PM, mb
L: 27. Small, fast-flying dove with *shorter tail and more pointed wings* than Eurasian Collared Dove. Told by *darker underwing, well-defined whitish belly-patch, rusty-edged, dark-spotted upperparts* with blue-grey outer wing-panel; also *contrasting pattern of uppertail with clear-cut white corners*, particularly when tail spread on landing. Sides of neck show *black-and-white-streaked patch in adult*, absent in juvenile. From Laughing Dove by build, jerky wingbeats, scalloped upperparts and tail pattern. The subspecies *arenicola*, breeding Near East to Iran and Arabia, *is paler, more washed-out grey-brown on mantle, wing-coverts (with less contrasting dark centres) and breast*. Often gregarious. **Voice** Soft, deep purring *roorrrr, roorrrr, roorrrr*, often persistent. **Habitat** Breeds in wooded country, sand desert with ghaf, oases; fields and livestock enclosures on passage. **Note** Passage hatched; rare in winter in S Arabia.

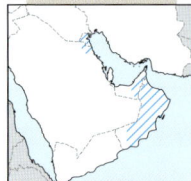

Oriental Turtle Dove *Streptopelia orientalis* pm, wv
L: 33. Resembles European Turtle Dove but *larger and heavier with broad-based wings, ill-defined dark centres to wing-coverts with narrow buff fringes* (broader tan fringes in European Turtle Dove); *forehead and crown pale grey, contrasting with browner rest of head*; neck-patch larger, *streaked blue and black* (usually white and black in European Turtle) and bare skin around eye *rounded* (larger area bare and lemon-shaped in European Turtle). Note also that Oriental Turtle Dove has broad buff fringes to primary coverts (narrow fringes in European Turtle Dove). Subspecies occurring in the region, *meena*, has whitish belly, undertail-coverts and distal part of tail. Juvenile lacks neck-patch, is paler brownish on body and wings, otherwise like adult. Flight heavier, straighter, with less jerky wingbeats than European Turtle Dove. **Voice** Alternating grating and clearer cooing notes, *gru-gror, co-co, gru-gror, co-co*. Migrants invariably silent. **Habitat** Open woodland, often near cultivation. **Note** Passage hatched, but rare; occasional in winter or even summer; vagrant Iran, Iraq, Saudi Arabia. [Alt: Rufous Turtle Dove]

PLATE 62: COLLARED, LAUGHING AND NAMAQUA DOVES

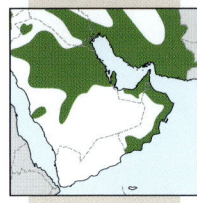

Eurasian Collared Dove *Streptopelia decaocto* RB, WV
L: 31–34. Medium-small, sandy-brown dove with narrow black half-collar on hindneck (absent in juvenile). Confusable with African Collared Dove, which see. *From European Turtle Dove by unspotted sandy-brown forewing* (dark spotting in Turtle), *neck collar, whitish underwing* (dark in European Turtle), plainer uppertail (bold pattern in European Turtle), *absence of well-defined whitish belly-patch*, grey undertail, longer tail and less rapid flight with *less jerky wingbeats and shorter, more rounded wings*. **Voice** Loud, deep, trisyllabic *coo-cooh-co* with stress on middle note, which is also drawn-out and highest in pitch (Barbary Dove stresses first note). **Habitat** Towns, villages, parks, fields; often in flocks. **Note** Ongoing range expansion in Arabia; some winter dispersal.

African Collared Dove *Streptopelia roseogrisea* rb
L: 29. Similar colour and pattern on upperwing, tail-corners and underwing as slightly larger Eurasian Collared Dove. Best separated by *white lower belly and undertail-coverts* (dirty grey in Eurasian Collared), rather shorter tail, more obvious eye-ring and *by voice*. *Dark border to entire trailing edge of pale underwing*. Young birds paler than adults, being whitish-grey on head and underparts (paler than Eurasian Collared). Gregarious, sometimes mixes with Eurasian Collared Dove. The domesticated Barbary Dove *S. risoria* is very like African and Eurasian Collared Doves but paler creamy, with clean white undertail-coverts, and lacks contrasting dark primaries above. **Voice** Distinctive, *high-pitched drawn-out note followed by a short pause then a series of broken, descending rolling notes*, *crooo, cro-cro-crococo or cruu… currruuu*; at distance sounds disyllabic *croo… cooorrr*. **Habitat** Semi-desert and savanna with trees; also mangroves, parks. **Note** Range expanding; probably introduced Kuwait; vagrant Bahrain.

Red Turtle Dove *Streptopelia tranquebarica* V
L: 23. Small, compact, short-tailed dove with black collar on hindneck. From larger Eurasian Collared Dove by combination of *red-brown* (male) *or warm-brown* (female) *mantle and wing-coverts*, *darker ash-grey lower back and uppertail with more defined white corners*, browner breast (vinous-brown in male), *whitish vent and undertail-coverts* (grey in Eurasian Collared) and *darker underwing* (pale in Eurasian Collared). Laughing and European Turtle Doves have darkish underwing but both lack hindneck collar. Fast flight like European Turtle Dove. **Voice** Dry, rattling, rhythmic *ruk-a-duc-doo*, quickly repeated. **Habitat** Open wooded country. **Note** Vagrant Iran, Oman, UAE.

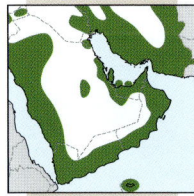

Laughing Dove *Spilopelia senegalensis* RB
L: 26. Small, dark red-brown to sandy grey-brown dove, with *black-spotted patch on foreneck and upper breast, unspotted red-brown upperparts and large, blue-grey area on outer wing-coverts*. Uppertail less contrasting than in European Turtle Dove. Juvenile lacks patch on foreneck and is duller. Population in SW Arabia is rather dark in plumage. Flight close to Eurasian Collared Dove but note short rounded wings and long tail. Gregarious; usually abundant. **Voice** Usually five syllables, subdued cooing with third and fourth notes slightly longer and higher in pitch, *do, do, dooh, dooh, do*. **Habitat** Towns, villages, gardens, oases and agricultural areas. **Note** Recent extensive range expansion in region. [Alt: Palm Dove]

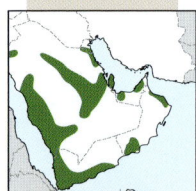

Namaqua Dove *Oena capensis* RB, PM, WV
L: 29 (including 9cm tail). *Very small, slim dove*, blue-grey or grey-brown with *long, pointed black central tail feathers*, recalling large Budgerigar. In flight, *black primaries show large red-brown patch*, rump with two transverse black bands. Male has black face and upper breast, which are brownish-grey in female. Juvenile barred black and buff on crown, throat, wing-coverts and back. Flight very fast and direct. Unobtrusive; spends much time on the ground. **Voice** Mournful *hu-hu, hu-hu*; also a deep coo. **Habitat** Savanna and semi-desert with thorn-bush or scrub. **Note** Range expanding in Arabia.

PLATE 63: GREEN PIGEON AND PARAKEETS

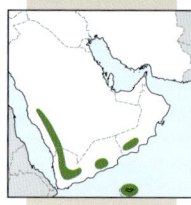

Bruce's Green Pigeon *Treron waalia* — RB, mb
L: 31. Unmistakable, brightly coloured dove with olive-green upperparts, *greyish-green head, neck and upper breast, sharply defined from bright yellow below*, undertail-coverts chestnut. Purple patch on shoulder is absent in juvenile. Uppertail uniform, undertail white with black base. *Flight very rapid*, wings making rattling sound. Hard to see when perched in tree and often shy; feeds in flocks. **Voice** Distinctive loud querulous crooning whistle somewhat like Tawny Owl, and a quarrelsome chatter. **Habitat** Open country, gardens with bushes and tall trees, especially figs, wooded wadis, palm groves. **Note** Partial migrant. In Oman, only in south. [Alt: Yellow-bellied Green Pigeon]

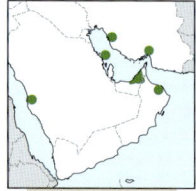

Alexandrine Parakeet *Psittacula eupatria* — E/I
L: 53–58. Resembles *distinctly larger version* of Rose-ringed Parakeet, with clearly heavier bill and *large red patch on shoulders at all ages* (sometimes partly hidden when perched), which is absent in Rose-ringed Parakeet. Rose-pink neck-ring of adult male easily seen; other sex and age differences as in Rose-ringed. **Voice** Range of notes uttered, most loud and hoarse, commonest a deep macaw-like screaming note *kerrrck* with stress on first syllable, repeated 2–3 times. **Habitat** Parks, gardens and plantations; nests in hole in tree or building. **Note** Not native; escaped birds now naturalised.

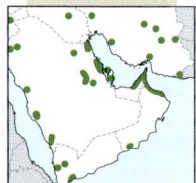

Rose-ringed Parakeet *Psittacula krameri* — RB
L: 42. Large green parakeet with *long, graduated, pointed tail and red bill*. In male, *black throat continues round neck as a narrow rosy ring*, absent in female; also absent in juvenile, which has a green throat and horn-coloured bill. Flight swift, fast and direct but flocks often change direction rather suddenly. **Voice** Noisy; loud screaming *kee-ak*, rather falcon-like and piercing. **Habitat** Gardens and open wooded; also near cultivation; nests, often colonially, in holes in tree or wall. **Note** Colonies outside Iran probably originate from escapes. [Alt: Ring-necked Parakeet]

Plum-headed Parakeet *Psittacula cyanocephala* — E/I
L: 36. Slim, fast-flying parakeet; tail blue-green with whitish tip. *Male has plum-red head; female has grey head with yellow upper breast and rear neck-collar*. Both sexes have yellow-horn upper and dark lower mandibles, as does plainer, green-headed juvenile. **Voice** A shrill *too-ik, too-ik*; male chatters in flight, often ending with distinctive, hurried ringing *der-wink*. **Habitat** Wooded parks and gardens. **Note** A few free-flying in Muscat area; may have bred but not yet naturalised.

PLATE 64: CUCKOOS I

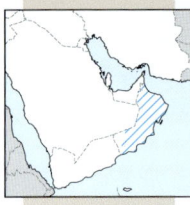

Asian Koel *Eudynamys scolopaceus*   pm, wv
L: 43. Large with stout bill, short wings and rather long tail. *Male black with bluish gloss and bright yellow-green bill.* Female has *drab brown upperparts, thinly streaked and spotted white*; throat heavily streaked dark brown, rest of underparts densely barred dark brown and buff; dark tail with many thin whitish bars. Immature resembles female. Eye garnet red. When not alarmed has characteristic hunched stance. Flight direct. **Voice** Seldom heard in region; loud whistled *koo-yl*; also a rapid ascending series of bubbling notes *kwow-kwow-kwow-kwow*. **Habitat** Scrub, gardens, woodland, fruit trees. **Note** Rare winter visitor to hatched area; vagrant Iran, Kuwait, UAE.

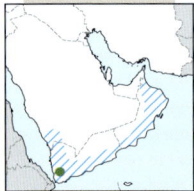

Jacobin Cuckoo *Clamator jacobinus*   pm
L: 33. Size of Common Cuckoo; distinctive *black-and-white plumage, crest, long graduated tail and conspicuous white wing-patches in flight*. A rarer black morph occurs (but which also shows white in wing). Juvenile is sooty above with dirty white underparts and smaller crest. **Voice** Loud, metallic *piu-piu-pee-pee-piu, pee-pee-piu*. **Habitat** Scrub, woodland edge. **Note** Scarce summer/passage visitor as hatched; vagrant Iran, UAE. [Alt: Pied Cuckoo]

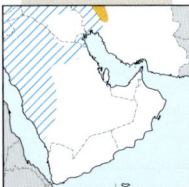

Great Spotted Cuckoo *Clamator glandarius*   V
L: 40. Large cuckoo with *slight crest, very long tail, grey upperparts and white-spotted coverts; underparts white*, with buffy throat. Juvenile is dark brown above with blackish head, white-spotted coverts and shows *distinctive chestnut primaries (above and below)*. Often on ground; flight rather fast, with shallow wingbeats. **Voice** Harsh, chattering *chil, chil, chil, chil*. **Habitat** Woodland, cultivation with bushes and trees. **Note** Passage hatched (rarer in autumn); vagrant Bahrain, Oman, Qatar, UAE, Yemen.

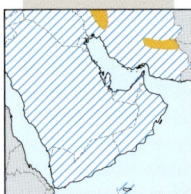

Common Cuckoo *Cuculus canorus*   PM, mb?
L: 33. In flight, has resemblance to Common Kestrel, but head held slightly raised and *wings not lifted above level of body*. Grey above with barring on underparts; grey underwing has pale line through centre. Female similar, with warm wash on breast. Rufous morph, in which female and juvenile have brownish-red plumage with noticeable barring on upperparts and breast, relatively common in Middle East. Grey morph juvenile is dark brownish-grey, with pale fringes above and whitish spot on nape. **Voice** Far-carrying song *unmistakable*, kwuk-koo or *kwuk, kwuk-koo*, first note highest; female has distinctive descending, long bubbling call. **Habitat** Open country, trees and bushes, *open* woodland. **Note** May breed Musandam/northern UAE; passage hatched.

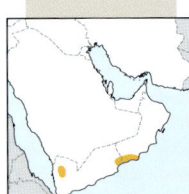

Diederik Cuckoo *Chrysococcyx caprius*   MB
L: 18. Slightly larger than Wryneck and of similar shape, with tail often held slightly raised. Metallic-green male and bronzy female told by *white spotting on wings, white supercilium, white crown-streak, whitish barring on outer tail feathers* and *bold barring on flanks, belly and undertail-coverts*. Juvenile told by plain rufous-buff upperparts and barred outer tail feathers. Rather secretive; appears quite long-tailed in flight. **Voice** Plaintive, loud *pee-pui-pui* (last note often double); *psee-psee-psu* (last note lower) or *dee-dee-dee-deederic* (emphasis on last note), often openly from high in tree. **Habitat** Dry scrub and open woodland; parasitises nests of Rüppell's Weaver in region. **Note** Breeding summer visitor to S Oman and Yemen; vagrant SW Saudi Arabia. [Alt: Didric Cuckoo]

PLATE 65: CUCKOOS II AND NIGHTJARS

Grey-bellied Cuckoo *Cocomantis passerinus*   V
L: 22. Rather small cuckoo. Male and most females easily told by *smooth, dark grey plumage fading into whitish belly and undertail-coverts*; wings have olive-brown wash; graduated tail has *white barring below*. In flight shows a conspicuous *white patch at base of underside of primaries*. Juvenile rufous-brown above, barred darker; chin and throat warm buff with fine black bars, rest of underparts white with narrow dark bars; undertail-feathers barred brown and white. Rather shy and when flushed often flies some distance. **Voice** Vagrants unlikely to call. **Habitat** Woodland, bushes and thickets. **Note** Vagrant Oman. Recently split from Plaintive Cuckoo *C. merulinus*.

Common Hawk-Cuckoo *Hierococcyx varius*   V
L: 33. Similar in size to Common Cuckoo, but in flight shows slightly *broader and shorter wings and slightly shorter tail*. In adult plumage grey above with *vinous wash on breast, faint barring on belly and banded tail (above and below)*. In juvenile (most likely to be seen in region), upperparts are dull rufous, barred darker with ill-defined white collar on hindneck of greyish head; underparts white with *dark streaks on throat and breast, and spotting on belly*. Rather secretive. In flight resembles a hawk, particularly Shikra. **Habitat** Usually in tall trees and scrub. **Note** Vagrant Oman. [Alt: Indian Hawk-Cuckoo]

European Nightjar *Caprimulgus europaeus*   PM
L: 26. The most widely encountered nightjar in the region but not resident. Distinguishing features (breeding subspecies *meridionalis* and passage migrant *europaeus*) are the *dark-streaked grey crown contrasting with browner cheeks and throat, grey upperparts with broad bands of dark brown and buff-white on the scapulars, and rows of whitish to buffy spotting on the coverts*. Male has white spots in wings and tail. Subspecies *unwini* (breeding in Iran and Iraq) is distinctly paler, with more sandy-grey upperparts, white-and-buff spotted coverts, and longer, whiter lower throat-patches; male shows larger white spots in primaries. (Migrant *plumipes*, assumed to occur, is paler sandy above, much as Egyptian Nightjar, but retains prominent white cheek-stripe and has greyish flight feathers). Most often seen in flight at dawn and dusk; will hunt moths attracted to street lighting or car headlights. Frequently encountered during day, perched lengthways along branch (also sits on ground) and if flushed usually flies only a short distance; note rather slow, soft wingbeats with long glides on stiffly held wings (as other nightjars), recalling cuckoo or falcon and similarly causing small birds in vicinity to alarm. Often sits on tracks at night. In courtship display, wing-claps in flight, often also giving distinctive *kru-ipp* call. **Voice** Song at night (only rarely during daylight) long rising and falling *churr*, which can go on for many minutes, alternating on two pitches and highly ventriloquial. Commonly calls again after dusk or pre-dawn, a distinctive, loud throaty *kru-ipp*, or falling, discontented guttural *kworr-kworr*. **Habitat** Edges of woods and heaths, steppes with sparse vegetation; any open areas on migration, often in or near trees or shrubs. **Note** Passage hatched.

Egyptian Nightjar *Caprimulgus aegyptius*   pm, wv
L: 25. The palest nightjar of the region, with a *pale sandy-grey ground colour, broad rows of inconspicuous buff tips to wing-coverts and white patch on side of neck (often very hard to see)*. Crown plain, finely dark-flecked on rear. *In flight, very pale underwing, and above, dark flight feathers contrasting with pale upperwing-coverts*. Both sexes lack white patch in wing, but male shows pale creamy spots on outer tail feathers below. Juvenile has ill-defined spotting, not barring, on underparts. From European Nightjar by pale sandy plumage, contrast between flight feathers and paler coverts and lack of white spots in less pointed wings. Invariably only ever sits on ground; often on tracks at night. **Voice** Song a regular, rapidly repeated *kowrr-kowrr-kowrr*, slowing towards end. **Habitat** Semi-deserts, often with palms or scrub. **Note** Passage hatched, often rare; winters in S Arabia.

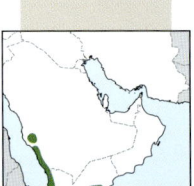

Nubian Nightjar *Caprimulgus nubicus*   V
L: 21. Slightly smaller than European Nightjar, with shorter tail. Told from other nightjars in the region by more rounded wings, *chestnut base to upper primaries, pale rufous underwing contrasting with black primaries* and conspicuous white patches on outer wing and outer tail feathers. When perched, helpful features are *buff half-collar, white bar on edge of cheek and another on side of throat*, and greyish upperparts with *broad buff tips to wing-coverts*. Wing-claps during display. **Voice** Song a fairly liquid *quil-quil* (recalling distant barking dog), repeated non-stop for up to 30 seconds, occasionally preceded by very quiet *poo-poo poo-poo*. **Habitat** Sand and stony deserts, including dry watercourses, with scattered vegetation such as palms and tamarisk. **Note** Vagrant Oman.

PLATE 66: BARN OWL AND SCOPS OWLS

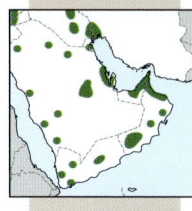

Western Barn Owl *Tyto alba* rb
L: 35. W: 89. Strikingly pale, with *creamy-buff upperparts, white underparts and heart-shaped face with black eyes*. Mainly crepuscular or nocturnal in region. **Voice** Territorial call of male is a clear, vibrant, chilling shriek of about two seconds; alarm call in flight is a shrill shriek; when disturbed will hiss; young beg with a drawn-out hissing. **Habitat** Open country with trees, semi-deserts, edges of woods, often near human habitation. Nests in hole in tree, building, ruins, crag, cave or nestbox. **Note** Much winter dispersal.

Pallid Scops Owl *Otus brucei* RB
L: 20–21. Typical scops owl, sandy-grey or grey above without white spots on crown or hindneck, usually with *clear-cut pencil-fine black streaks below* (grey morph of Eurasian Scops Owl has black streaks crossed by vermiculations and paler blotches below, submerged by darker ground-colour). Braces creamy to cinnamon-ginger. *Tail extends marginally beyond wing-tips; tail-bands usually diffuse, especially towards tip of tail. Feathering on legs extends to base of toes, middle toe especially. Juvenile completely barred below* (unlike juvenile Eurasian Scops). **Voice** Diagnostic, *soft and dove-like*, carrying only a short distance (unlike Eurasian Scops): hollow, resonant, low-pitched *whoop* or *whoo* repeated regularly about eight times in five seconds (like distant water-pump); also longer *whooo* repeated irregularly at 3–5-second intervals or *ooo-ooo... ooo-ooo*. **Habitat** Arid hills, semi-desert, wadis with trees, parks, palm groves. Nests in hole. **Note** Passage and winter hatched; vagrant Bahrain, Kuwait, Qatar, Yemen. [Alt: Striated Scops Owl, Bruce's Scops Owl]

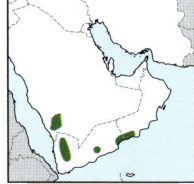

Arabian Scops Owl *Otus pamelae* RB
L: 19. Similar to Eurasian Scops Owl and *best told by voice*. Shows the most *prominent pale eyebrows* of the scops owls in the region; facial disc pale with dark rim; upperwing-coverts have rufous tinge and *finely barred greyish underparts have darker streaks*. Leg feathering does not reach base of toes. In the hand, the outermost primary, P10, shorter than fifth or fourth (less than fifth in Eurasian Scops). **Voice** Single *da-pwoorp* repeated at 6–10-second intervals or more. **Habitat** Wooded hills to higher elevations. Nests in hole in tree. **Note** Resident in S Oman. Formerly considered a subspecies of African Scops Owl *Otus senegalensis*.

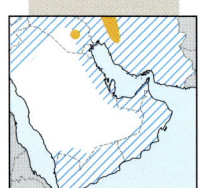

Eurasian Scops Owl *Otus scops* PM
L: 18–20. Typical scops owl with small ear-tufts; *best distinguished by voice* though migrant birds are usually silent. Mothy grey-brown, or rufous-brown (rarer morph), with paler face and dark surround to yellow eyes. *Wing-tips reach tail-tip; whitish tail-bands bordered narrowly with black and usually clear-cut* (but often cloaked by closed wings). Upperparts streaked and vermiculated; braces off-white, buff or creamy, sometimes rusty on outer webs, with parallel *deep rust or rufous line outside braces* (if not hidden); underparts streaked, barred and vermiculated, *interspersed with white blotches* (lacking in Pallid Scops). Feathering on legs stops square-cut, short of toe bases. Nocturnal **Voice** Clear, soft *whistle repeated rhythmically* and monotonously, *pwoo, pwoo* at 2–3-second intervals. **Habitat** Trees, groves. **Note** Passage hatched; occasional in winter.

PLATE 67: LARGER OWLS

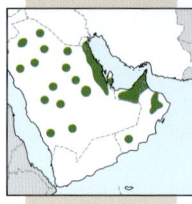

Pharaoh Eagle-Owl *Bubo ascalaphus* rb
L: 50. A large owl, *cinnamon or ginger in tone*, with most of *lower breast and belly distinctly barred*. Eyes orange-yellow. Often sits out in daytime. **Voice** Loud; a booming *boooor* and variety of other, mainly deep calls. **Habitat** Mountains, steppe, frequenting cliffs, crags and rocky outcrops; sand desert with large trees or bushes. Nests on ledge, down well, under roots or in old tree nest of raven or buzzard. **Note** Resident with some dispersal post-breeding and wandering in winter; vagrant Bahrain. [Alt: Desert Eagle-Owl]

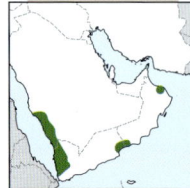

Spotted Eagle-Owl *Bubo africanus* RB
L: 45. Endemic Arabian subspecies *B. a. milesi* (Arabian Spotted Eagle-Owl) occurs. Resembles Pharaoh Eagle-Owl, but *entire underparts including underwing-coverts densely barred dark brown* with only irregularly dark-blotched upper breast; upperparts barred dark brown with white spots on edge of scapulars; paler Pharaoh Eagle-Owl has streaked breast and centre of belly often unbarred, others have dark shaft-streaks in the barring (absent in Arabian Spotted Eagle-Owl); *nape and hindneck finely white-spotted* (streaked in other eagle-owls). In flight, shows bolder pale and dark tail bands than in other eagle-owls. Eyes yellow. Often active during day. **Voice** Male has a double-note song *hu-hoo*, often in duet with lower-pitched female's rather dove-like *hoo-doo-doh-dooh* (second note highest, last two lowest). **Habitat** Open woodland, rocky hills, ravines; sometimes near habitation. Nests on ground under rock, cliff ledge or occasionally in tree. **Note** Rare breeding resident in N and S Oman.

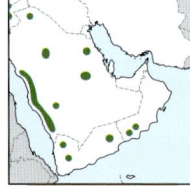

Desert Owl *Strix hadorami* rb
L: 37. W: 96. *Like a pale Omani Owl with yellowish-orange eyes*; occurs in desert regions. *Facial disc off-white with dark band in centre of forehead*; underparts whitish-buff with indistinct ochre barring; upperparts sandy- or greyish-brown with bold barred uppertail and pale golden-buff collar. Strictly nocturnal. **Voice** Territorial call a five-syllabic *whoooo, hoo-hoo, who-who* with stress on first drawn-out note. **Habitat** Rocky deserts with gorges, desert earth-banks; often near palm groves, acacias and sometimes near springs and settlements. Nests in hole in rocks or cliff face. **Note** Resident in much of Arabia including S Oman. Formerly known as Hume's Owl *Strix butleri*.

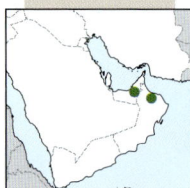

Omani Owl *Strix butleri* rb
L: 35. Very similar to Desert Owl and lacking ear tufts, but *darker and less rufous*. Bicoloured facial disc, pale below and between the eyes, darker above the eyes. Facial disk surrounded by pale orange-brown ring. Orange eyes like Desert Owl. Upperparts dark greyish-brown. Breast pale with short vertical stripes becoming whiter on belly. In flight, upperwing greyish-brown with some whitish spots; underwing pale with darker tips to secondaries and primaries. **Voice** Four syllabic *hoo-HOO-hoo-hoo* with stress on second note and short third syllable. Also a pulsed hooting of almost identical notes. **Habitat** Steep-sided mountain wadis. **Note** Resident in Iran, N Oman and eastern UAE.

PLATE 68: SMALLER AND MEDIUM-SIZED OWLS

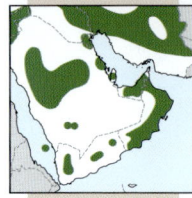

Little Owl *Athene noctua* RB
L: 22. Small with *round, flat-crowned head and long legs*. Subspecies vary in upperparts colour. Pale Arabian birds usually assigned to the subspecies *saharae*, but birds in eastern Arabia often cold chocolate-brown and may represent another subspecies. An even paler form (considered by some to be of the subspecies *lilith*) may also occur and this is illustrated. All subspecies have *crown and nape distinctly white-spotted*, with white-blotched upperparts and wings, *underparts boldly streaked*; eyes yellow, framed white. Flight deeply undulating, alternating rapid flapping with closing of wings. Largely crepuscular, but often sits in the open in daytime on rock, building or telegraph pole. If agitated, may bob (pogo) in an upright posture. **Voice** Territorial call is a drawn-out and wailing *koooah*, also a short *kiu*; alarm call is a sharp series of dog-like yapping notes *kip-kip-kip...*. **Habitat** Open country with trees, stony wasteland, wadis, rocky semi-deserts, sand desert with outcrops, cultivated areas. Nests in hole in tree, in rocks, buildings and burrows. **Note** The subspecies *lilith* is sometimes considered to be a separate species, Lilith Owl *A. lilith*.

Long-eared Owl *Asio otus* V
L: 36. W: 95. At roost, muted brown and buff with distinctive *long ear-tufts* (though invisible when flattened); *facial disc noticeably warm buff with striking white divide and orange eyes*. Flight jerky; fairly stiff wingbeats with glides on level wings (sometimes slightly raised). Separated from Short-eared Owl mainly by face pattern, *entirely streaked underparts* and wing pattern in flight (see Short-eared Owl). Nocturnal. **Voice** Vagrants silent. **Habitat** Woodland. **Note** Winter hatched; vagrant Kuwait, Oman, Qatar, Saudi Arabia, UAE.

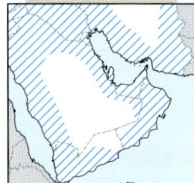

Short-eared Owl *Asio flammeus* WV
L: 38. W: 102. Often diurnal, flying with slow, *elegant and high wingbeats* on long, slender wings, *raised during glides in shallow 'V'* (shorter wings with shallower, faster wingbeats in Long-eared); sits on ground. Further differs from Long-eared Owl in strongly buff-spotted upperparts, paler greyish facial disc with striking black surround to *glaring yellow (not orange) eyes, ear tufts short*. In flight, separated from Long-eared by *yellow-buff base of primaries* (buff-orange in Long-eared), *more contrasting bars on flight feathers and tail, white trailing edge to upperwing*, black tips to underwing (not diffusely barred, as in Long-eared Owl) and *dark streaking on underparts mostly confined to breast and contrasting with paler belly*. **Voice** Silent in winter. **Habitat** Open country, often marshy. **Note** Winter and passage hatched; rare in S Arabia; vagrant Qatar.

Little Owl

lilith

Long-eared Owl

Short-eared Owl

PLATE 69: SWIFTS

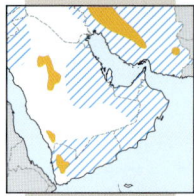

Alpine Swift *Tachymarptis melba* — pm, wv
L: 21. W: 57. Shape and behaviour as Common and Pallid Swifts, but *much larger with markedly slower and deeper, scything wingbeats; white underparts broken by dark breast-band; vent and undertail-coverts brown*, as upperparts. **Voice** Loud, dry chattering trill unlike that of Common Swift, *trit-it-it-it-itititit-it-it-it*, accelerating then decelerating, rising and falling. **Habitat** Mountain ridges, steep rock faces, also sea cliffs. Nests, usually colonially, in natural crevices on cliff face, occasionally in ancient building. **Note** Mainly summer visitor; passage hatched, rare SE Arabia; winter records Iran, UAE (where vagrant), Yemen.

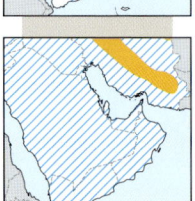

Common Swift *Apus apus* — pm
L: 16. W: 45. *Almost uniform sooty in fresh plumage*, but brownish and more contrasting in worn plumage and in eastern subspecies *pekinensis (which migrates through Oman)*, though much depends on light conditions. *Whitish round throat-patch, but variable; lacks contrast above but underparts show inner flight feathers clearly paler than body and wing-coverts*. Easily confused with Pallid Swift (which see). Wing moult of adult is entirely in winter quarters (unlike Pallid Swift). **Voice** High-pitched ringing *shreee*, similar to Pallid Swift. Mostly silent on migration. **Habitat** Aerial; congregates in areas with suitable nesting sites or abundant food. Nests in buildings, under eaves, occasionally on cliffs. **Note** Migrant breeder; passage hatched.

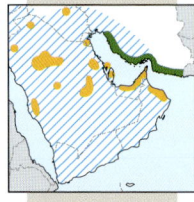

Pallid Swift *Apus pallidus* — PM, MB
L: 16. W: 44. Slightly bulkier than Common Swift but all brownish with broader, less tapering outer wing and sometimes blunter wing-tips; tail-fork slightly shallower (adult) with less pointed tips. Often difficult to separate from Common Swift but *head looks broad and flat with larger, more triangular, whitish throat-patch; forehead and lores paler, contrasting with dark eye-patch. Faint scaling visible on underparts; upperparts show darker 'saddle'*, slightly contrasting with paler inner flight feathers, head and rump. Underwing shows darker outer primaries than inner wing, much as in Common Swift. Adults commence wing moult in breeding areas in summer/autumn, unlike Common Swift, which often appears badly worn. **Voice** Similar to Common Swift, but deeper, hoarser (tinnier, less ringing; individual notes less clear-cut) and slightly disyllabic, *sree-er*. **Habitat** Nests in historic buildings, new towers, cliffs or craggy outcrops. **Note** Migrant breeder; passage hatched.

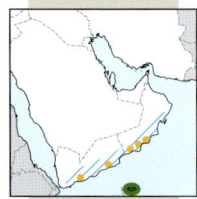

Forbes-Watson's Swift *Apus berliozi* — MB
L: 17. W: 46. Separation from Common and Pallid Swifts extremely difficult. Plumage blackish-brown with *distinct slightly pointed white chin and throat* (rounded in Pallid Swift) and slight white forehead. At close range *narrow pale feather fringes just detectable on upperparts, belly and undertail-coverts*; upperwing can occasionally show an oily, *greenish sheen on outer primaries, secondaries and median coverts*; the underwing has paler, almost translucent secondaries and inner primaries. Flight similar to Common Swift; wingbeats perhaps slower, and with longer, sweeping glides. In the hand, outermost (tenth) primary longest (ninth primary longest in Common Swift). From Pallid Swift by less noticeable saddle, slightly darker plumage, oily sheen to feathers on wing, and voice. **Voice** A dry screeching *schweee* or *schweee-weee-eee*, less rippling and not as high-pitched as Common Swift. **Habitat** Mountains, sea cliffs, foothills and plains. **Note** Mainly summer visitor to S Arabia; known passage/dispersal hatched, but much confusion due to difficulty of identification.

Pacific Swift *Apus pacificus* — V
L: 18–19. Larger than Common Swift, with *white rump, more deeply forked tail*, longer wings, *pale scaling to underparts* and slimmer build with *attenuated (Arctic Skua-like) rear end*. Also *more protruding head* and larger white chin-patch. **Voice** Unlikely to be heard in region; coarser and harsher than Common Swift, closer to that of Pallid Swift and dropping off at end. **Habitat** Open country. **Note** Vagrant Oman, UAE. [Alt: Fork-tailed Swift]

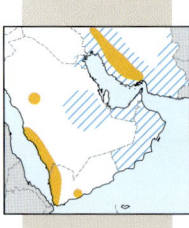

Little Swift *Apus affinis* — pm
L: 12. W: 34. Smallest swift in the region, distinctive stout silhouette with *short, square-ended tail* (round when spread); prominent *deep, white rump-band 'wrapped-around' onto rear flanks* (some white visible at all angles). At longer distance can recall house martin, but dark underparts and stiff wingbeats separate. Somewhat fluttering flight, alternating with short glides. **Voice** Fast, high-pitched, rippling trill *dillillillillill*, regularly rising and falling in pitch, much higher-pitched and faster than Common Swift. **Habitat** Over grassland or near water, often with other swifts; also gorges, towns and cities (nests colonially on ceiling of open building, under rock overhangs, cave roofs). **Note** Passage hatched, but rare; vagrant Bahrain.

PLATE 70: ROLLERS, HOOPOE AND WRYNECK

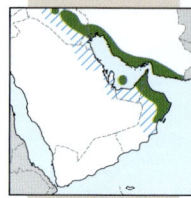

Indian Roller *Coracias benghalensis* RB
L: 30. Stocky and multicoloured; told at all ages in flight *by large, pale turquoise-blue primary patch* (above and below) *and pale turquoise-blue rectangles in sides of tail base* (seen when tail is spread). Wing-tip clearly blunter than in European Roller. When perched, *lightly white-streaked neck, throat and breast are vinous-cinnamon* (turquoise-blue in European Roller) *and cap dark turquoise-green*; mantle earth-brown (pale chestnut in European). Aerobatic, In pursuit of flying insects and in sky-diving display. **Voice** Similar to European Roller, but with more barking *rak*; agitated sneezed *chew-chew-chew* in display or towards intruder in territory. **Habitat** Open cultivated country with scattered trees, plantations, parks, gardens; usually below 1,000m. Nests in hole in tree or wall. **Note** Mainly resident, autumn dispersal hatched, but rare; vagrant Qatar, Yemen.

Lilac-breasted Roller *Coracias caudatus* V
L: 41. Length includes *elongated outer tail feathers*. Superficially like European and Indian Rollers but with long tail streamers; distinguished by *rich lilac throat and sides of face* in the Somali race *lorti* (Lilac-throated Roller), the race most likely to occur. Upperparts earth-brown, similar to Indian Roller, but wing pattern more similar to European Roller. Note *whitish forehead and supercilium*. In juvenile and moulting adult, lilac throat is duller and browner. Behaviour as other rollers. **Voice** Harsh loud *krack-krack*; also chattering *kark*. In flight, a sharp rasping *kick-kick-kick*. **Habitat** Plains with trees. **Note** Vagrant Oman, Yemen.

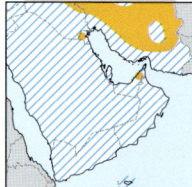

European Roller *Coracias garrulus* PM
L: 30. Stocky; *turquoise-blue body and most wing-coverts contrasting with blackish flight feathers, chestnut back and deep blue leading forewing above*. Colours often faded in autumn (and paler-hued in subspecies *semenowi*, breeding in Iran and Iraq). Juvenile duller and browner with lightly streaked neck and breast. Often sits on prominent perch (wires, poles, dead branches) taking prey on ground. In display flight, 'tumbles' from side to side in downward dive (not unlike Northern Lapwing). **Voice** Sonorous, hoarse *rack-rack* (recalling Eurasian Magpie); also in display or when agitated a loud piercing *keer-keer-keer* and repeated dry gravelly grating note. **Habitat** Open country with large trees, rarely to 2,000m. **Note** Passage hatched; formerly bred UAE.

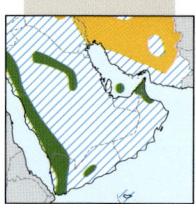

Eurasian Hoopoe *Upupa epops* WV, PM, rb
L: 28. *Distinctive pinkish-buff with bold black-and-white bars on wings and tail, long black-tipped crest, usually depressed* (raised on landing), *long decurved bill; broad rounded wings in flight* and flappy flight action. Spends much time on the ground. **Voice** Male's song distinctive, repeated hollow *poo-poo-poo*; also a dry *terrr* when agitated and strange, thin squeaking and hissing notes when courting. **Habitat** Woodland, olive and palm groves, parks, gardens, oases; open and wooded areas in winter. Nests in hole in tree or ruin. **Note** Passage and winter hatched.

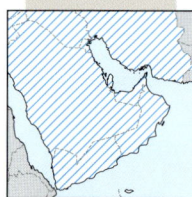

Eurasian Wryneck *Jynx torquilla* PM, wv
L: 16. Inconspicuous slim, atypical, brown woodpecker with finely vermiculated cryptic plumage. Flight direct, shallowly undulating, reminiscent of large, long-tailed warbler or female Red-backed Shrike. Distinctive features are long tail, *brown scaling on off-white underparts, dark eye-stripe, black central crown streak onto mantle, buff-mottled wings and finely barred yellowish throat*. Often forages on ground. **Voice** Song a loud, monotonous, plaintive *vee-vee-vee-vee* recalling small, distant falcon. **Habitat** Open woodland, orchards, parks; nests in hole in tree. Any cover on migration. **Note** May breed NW Iran/NE Iraq; passage hatched, some winter in Arabia.

PLATE 71: KINGFISHERS

White-throated Kingfisher *Halcyon smyrnensis*   V
L: 26. Large, brightly coloured kingfisher with *enormous red bill, dark chocolate-brown head and belly, large white throat and breast 'bib', brilliant turquoise-blue upperparts* and black forewing. Flight fast and straight, showing conspicuous white primary patches. Often perches on wires (but can sit hidden) looking for prey on ground. Rather noisy. **Voice** Loud raucous yelping *kril-kril-kril-kril*; also tittering descending song. **Habitat** Dry woodland glades and palm groves, as well as tree-lined lakes, rivers or other wetlands. **Note** Some winter dispersal, including to Saudi Arabia; vagrant Oman, Qatar, UAE.

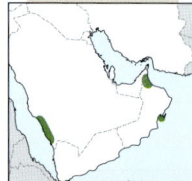

Grey-headed Kingfisher *Halcyon leucocephala*   MB
L: 20. Fairly large, with large red bill, *buffish-grey head and neck, off-white throat and upper breast; bluish-black wing-coverts and back, with chestnut belly*. White patch in primaries visible in flight. In first autumn, slightly darker head, dark scalloped pectoral band and black-tipped bill. Often sits on dead branch, poles, wires or other prominent perch. Not dependent on water. Flight sluggish and undulating. **Voice** Weak chattering *ji, ji, ji-jeee*. **Habitat** Foothills, wadis with trees (with or without water). **Note** Summer visitor to S Oman. Vagrant UAE.

Collared Kingfisher *Todiramphus chloris*   rb
L: 24. Large kingfisher easily told by *turquoise upperparts, all-white underparts and white collar*, bordered above by long black eye-stripe and white supercilium to nape; base of bill horn; upper mandible greyish. Juvenile has dusky barring on breast. Flight appears rather weak. Noisy, particularly at dawn, intermittently at other times. Often perches low in mangroves on look-out for crabs, on which it mostly feeds; when perched on ground tail usually held cocked. **Voice** Distinctive kookaburra-like calls; a fast series of loud, ringing notes, each ascending, until end *chei-chei-chei-chei*. **Habitat** Mangroves. Nests in hole in tree or bank. **Note** Endangered endemic subspecies *kalbaensis* restricted to Khor Kalba, Sharjah, and nearby sites on Batinah of Oman; also resident on Mahawt Island off east coast of Oman. [Alt: Mangrove Kingfisher]

Malachite Kingfisher *Corythornis cristatus*   pm, wv, rb?
L: 12. Smaller than Common Kingfisher, which it otherwise resembles, but upperparts darker blue and *red underparts reach to eye*, not broken by broad bluish stripe as in Common Kingfisher. *Dark-barred crown sometimes raised as shaggy crest*. Bill red in adult (largely black in Common Kingfisher), though blackish in juvenile. Behaviour much as Common Kingfisher. **Voice** Sharp, not very loud, *teep-teep*. **Habitat** Permanent water with vegetation. **Note** Rare visitor to S Oman, Yemen; breeding attempted in S Oman.

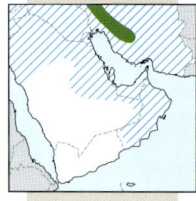

Common Kingfisher *Alcedo atthis*   PM, WV
L: 17. Small kingfisher with *brilliant blue and green upperparts*, and *reddish-orange underparts* (often bleached in autumn), with buffish-white throat and neck-patch, and long, blackish bill (though female has red on most of lower mandible). In flight, which is swift, direct and low over water, luminous back and tail are obvious. Fairly shy, often inconspicuous when perched, sitting on overhanging branch for long periods before diving for fish; often hovers. **Voice** High-pitched, thin, piercing *cheee* or *tzee*, mostly in flight. **Habitat** Rivers, streams, canals, lakes; in winter also coasts. **Note** Passage and winter hatched; vagrant Yemen.

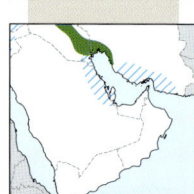

Pied Kingfisher *Ceryle rudis*   V
L: 25. Large and unmistakable; *the only black-and-white kingfisher in region, frequently seen hovering well above water*, plunging for fish (but also fishes from perch). White underparts have two more or less complete black breast-bands in male, one in female. Black eye-mask, white supercilium, short crest on nape; white-sided tail has black band at tip. Juvenile has greyish breast-band. Sometimes in small groups. Rather vocal. **Voice** Loud, noisy chattering *chirrick, chirrick, chirrick*. **Habitat** Rivers, lakes, ponds, mangroves and open coasts. Nests in hole in bank. **Note** Winter hatched; vagrant Oman, UAE.

PLATE 72: BEE-EATERS AND TCHAGRA

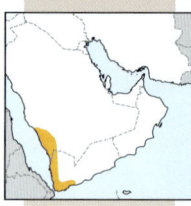

White-throated Bee-eater *Merops albicollis* V

L: 30. More elegant than European Bee-eater, *easily distinguished by black-and-white head pattern, white underparts with black collar around throat and very long central tail streamers*; also has bluish-green upperparts and blue tail. In flight shows ochre upperwing and coppery underwing, both with black trailing edge. Often remains below tree canopy and easily overlooked. **Voice** Higher-pitched and softer than European Bee-eater, *prrrp, prrrp, pruik*. **Habitat** Hills, plains and wadis with bushes and trees; also agricultural land. **Note** Vagrant Oman, UAE.

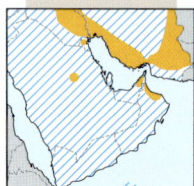

Blue-cheeked Bee-eater *Merops persicus* PM, MB

L: 30. W: 48. Larger than European Bee-eater and *distinctly green or turquoise-green* with long central tail-streamers and strongly *rusty-red underwings* framed with dark line along trailing edge. Juvenile duller and lacks long tail projections. From young European Bee-eater by entirely green plumage, including crown and underparts, and rusty-red underwings. Gregarious; hunts insects in flight. Vocal and audible at long range. **Voice** Very similar to European Bee-eater, but higher pitched and *certainly hoarser (throatier)*, the notes sometimes disyllabic, *prrllip-prrllip* or *prl-rip*. **Habitat** Dry open country with scattered trees; often perches on overhead wires. Almost anywhere on diurnal passage. Nests colonially in holes excavated in sandy ground. **Note** Passage hatched.

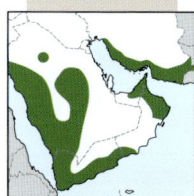

Green Bee-eater *Merops orientalis* RB

L: 24. W: 30. *Small, mainly green bee-eater* with black eye-stripe and elongated central tail feathers. Arabian birds (*muscatensis* in N Oman, *cyanophrys* in S Oman) have blue on supercilium and throat, rather diffuse dark breast-band (often absent) and shortish tail-streamers. Crown and nape shining coppery-gold in fresh-plumaged adults. Usually in pairs. **Voice** In flight, a high-pitched *treet-treet* or *prrrit*; often burbles excitedly. **Habitat** Open country with trees, semi-desert, wadis, cultivations, parks, gardens. Nests in tunnel in bank or hole excavated in ground. **Note** Some dispersal and seasonal movements; vagrant Bahrain, Qatar. [Alt: Little Green Bee-eater]

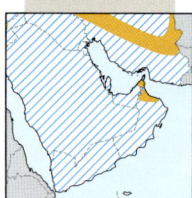

European Bee-eater *Merops apiaster* pm, mb

L: 28. W: 46. Easily distinguished by *chestnut crown and back, bright yellow throat contrasting with turquoise-blue underparts* and, in adult, *chestnut upperwing-coverts*. In flight shows paler, pinker underwings than other bee-eaters in the region. Juvenile has greenish upperparts, but shows chestnut crown and yellow throat (both paler); central tail feathers mere spikes or lacking. Migrates in vocal flocks, often high overhead, sometimes audible but remaining unseen. Hunts insects in flight. **Voice** Similar to Blue-cheeked Bee-eater but softer, lower pitched and more liquid, a far-carrying *pruup*, usually not disyllabic. **Habitat** Open bushy country with scattered trees, riversides and woodland glades; often on overhead wires. Nests colonially in holes excavated in sand banks, riversides and roadside cuttings. **Note** Declining breeder in N Oman; passage hatched.

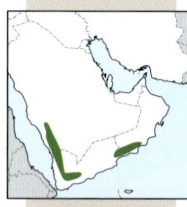

Black-crowned Tchagra *Tchagra senegalus* RB

L: 22. Thrush-sized with long, dark, conspicuously white-tipped tail, striking head pattern, large chestnut wing-patch and heavy black, slightly hooked bill. Secretive with clumsy movements. Cocks and flicks tail; will hop on ground with tail raised, or glide from bush to bush. Often in pairs. **Voice** Song mainly early morning or evening, hidden or in songflight; fluty and melodious, *c.* 10 notes with second part descending. Rises with soft wing-noise, singing on glide down from top of arc. Also characteristic long rising and falling trill, *truit-truit-driririririvivir*; alarm a harsh *shrrr*. **Habitat** Dense dry scrub with scattered trees. **Note** Breeding resident in S Oman. [Alt: Black-crowned Bush-shrike]

PLATE 73: SHRIKES I

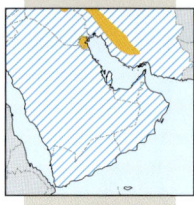

Woodchat Shrike *Lanius senator* pm, mb?
L: 18. Perches prominently; *rufous-brown crown and nape and large white shoulder-patch on black upperparts* diagnostic. Female duller than male, with more white-buff at base of bill, and rump only slightly greyer than mantle. Juvenile/first-winter resembles juvenile Red-backed Shrike but has *paler grey-brown upperparts with creamy-white markings on scapulars, wing-coverts, rump and base of primaries* (pale wing-patch noticeable in flight). **Voice** Song comprises varied whistles and trills, often mimics; calls *schak-schak* and grating rattles. **Habitat** Open bushy country, parks. **Note** Passage hatched; rarely winters Arabia and Iran.

Masked Shrike *Lanius nubicus* pm, wv
L: 18. Male with *black upperparts and mask, white face, orange wash on flanks* (female similar but duller) and distinctive white shoulder-patch. *White base of primaries and outer tail feathers obvious in flight*. Juvenile similar to juvenile Woodchat but with finer bill, longer tail, paler forehead, colder, *greyer ground colour to upperparts and rump concolorous with back*. First-winter lacks orange flanks. Perches mid-level in tree or bush with *long tail pumped slowly downwards or slightly cocked and waved up and down*. Sometimes secretive. **Voice** Harsh, scolding, dry *krrrrrr*. Song rather quiet, a tuneless jumble of notes. **Habitat** Open woodland, cultivation, parks, scrub. **Note** Passage hatched; occasional in winter in S Arabia.

Brown Shrike *Lanius cristatus* V
L: 18–19. Dark rufous-brown to rusty olive-brown upperparts; tertials pale-fringed, *crown, mantle and tail concolorous; tail graduated*, sometimes faintly barred. Subspecies *luscionensis* has grey crown and nape; intergrades with nominate *cristatus* occur. Breast creamy- to yellowish-buff, throat white, flanks rufous-buff. Bill usually appears deep, heavy. Immature told from similar immature Red-backed Shrike by *absence of white at base of primaries* (rarely small amount present), *shorter primary projection (only 4–5 primary tips showing beyond tertials), tail shape and underparts colour*, bill pale-based with dark tip. Often furtive in cover. **Voice** Loud, excitable staccato ratcheting. **Habitat** Scrub, woodland edge. **Note** Vagrant Oman, UAE.

Red-backed Shrike *Lanius collurio* PM
L: 18. Adult male unmistakable; female and first-autumn birds warm brown above, tail dark brown, underparts whitish with dense crescentic barring on breast and flanks, also above in immatures. Barring, colour of rump, tail and upperparts separates from Isabelline and Red-tailed Shrikes. Immature similar to Woodchat Shrike (which see). *Primary projection long*, thus eliminating Brown Shrike. **Voice** Song quiet and musical warbling with mimicry. Call a short *shack*, alarm hoarse, hard *keck-keck-keck*. **Habitat** Scrub, thickets, lightly wooded areas. **Note** Passage hatched.

Isabelline Shrike *Lanius isabellinus* PM, WV
L: 16–18. Subspecies *isabellinus* occurs. Pale sandy to sandy-grey above, whitish to creamy- or orange-buff below, with burnt orange flanks. *Rump and tail foxy- to orange-red*, or dull brown in some females, square-ended or only slightly rounded. Immature has *upperparts except mantle finely barred* (crescents on upperparts of Red-backed Shrike have *dark and pale fringes adjacent*); faintly barred on flanks. In female, white at base of primaries often reduced, rarely absent. Adult from Red-tailed Shrike by *paler upperparts, creamy to warm orange-buff supercilium (never white), pale lores and creamier breast and belly*. **Voice** Short, hard dry staccato ratcheting. **Habitat** Open wooded or scrubby areas, cultivations, parks. **Note** Passage and winter hatched [Alt: Daurian Shrike]

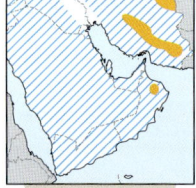

Red-tailed Shrike *Lanius phoenicuroides* PM, mb?
L: 16–18. Sandy grey-brown to rich warm earth-brown above, some almost as rich as Brown Shrike, with contrasting foxy-red rump and tail, latter often ruddier-chestnut than Isabelline Shrike. Spring male from Isabelline by *rufous crown, white supercilium, darker upperparts and dark lores, also whiter underparts* with flanks rusty; female and immature similarly darker above but probably not all immatures safely distinguishable. Intergrades occur; form *karelini* sandy or sandy-grey above with pale rufous to concolorous sandy-grey crown. From similar Brown Shrike by tail shape, slightly longer primary projection (six primaries exposed beyond tertials, but beware in winter moult) and white at base of primaries (sometimes not visible at rest). **Voice** Calls as Isabelline Shrike. **Habitat** As Isabelline Shrike. **Note** Passage hatched; rare or absent in winter. [Alt: Turkestan Shrike]

PLATE 74: SHRIKES II

Bay-backed Shrike *Lanius vittatus*   pm
L: 18. Plump, long-tailed and large-headed shrike; adult has *broad black forehead-band reaching forecrown, greyish rump, whitish upper tail-coverts, white sides to tail and large white mirror at base of primaries, forming conspicuous bar in flight* (larger Long-tailed Shrike has graduated tail with buff outer feathers, narrow black forehead-band, bright orange-buff rump and upper tail-coverts, and minute wing-mirror; female duller than male. Juvenile has rufous tail and faint wing-mirror, told from young Isabelline and Red-tailed Shrikes by rufous-edged greater coverts and grey lower rump. First-winter birds superficially like adult but have wavy barring on flanks and upperparts, and the black forehead-band may be absent. **Voice** Calls grating; song quiet and pleasant, imitating other species. **Habitat** Open, often rocky country with scattered trees and scrub. **Note** Rare passage migrant Oman (has bred); vagrant Qatar, UAE (has bred).

Lesser Grey Shrike *Lanius minor*   pm
L: 20. Resembles Southern Grey Shrike but smaller with *stouter bill, proportionately longer wings with long primary projection* and shorter, less graduated tail; *broader white wing panel* confined to primaries. *Adult shows extensive black forehead* without white supercilium, bluish-grey upperparts lacking white on shoulders (sometimes faintly), *and pinkish-white underparts*. Juvenile/first-winter lack black forehead, have paler bill, brown-grey upperparts, finely barred darker, and paler tips to wing-coverts and flight feathers; underparts creamy-white, sometimes faintly barred. **Voice** Chattering, varied whistles, trills and mimicry. **Habitat** Open cultivated country with scattered trees and bushes. **Note** Passage hatched.

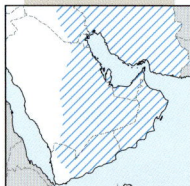

Steppe Grey Shrike *Lanius pallidirostris*   PM, WV
L: 25. Much as Southern Grey Shrike with very pale grey upperparts, white underparts usually blushed pinkish-buff on breast and flanks, *pale (not black) lores, broad white wing-bar not extending onto secondaries*; in winter, *bill horn-coloured or grey*, tipped dark (but all black in summer). **Voice** Demonstrative raucous screeches, mechanical throaty whirrs and clicks strung together. **Habitat** Steppe and semi-desert scrub. In winter *sparsely vegetated habitats*, mostly sand desert and semi-desert scrub, acacia savanna. **Note** Passage and winter hatched.

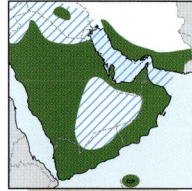

Southern Grey Shrike *Lanius meridionalis*   RB
L: 25. Resembles Steppe Grey Shrike closely, but adult *lacks white on secondaries*. Juvenile pale-billed, without black over base of bill, unbarred above (unlike juvenile Lesser Grey); told from wintering adult Steppe Grey by ochre-buff on breast and wing-bars. Several subspecies occur in region; *aucheri* (resident Oman, UAE) is dark grey above with grey wash on flanks, dark lores, broad black line over bill and relatively small amount of white in wing. **Voice** Song is a mixture of quiet, melodious ramblings, raucous notes and mimicry. Calls includes harsh *sheck sheck*, often extended into chatter. **Habitat** Open wooded and scrubby areas, rarely such sparse habitats preferred by Steppe Grey. **Note** Winter hatched.

Long-tailed Shrike *Lanius schach*   pm
L: 24. Size of Southern Grey Shrike, but *graduated tail with buff (not white) outer feathers, and rump and upper tail-coverts bright orange-buff*. Subspecies *erythronotus* depicted. White mirror at base of primaries small and inconspicuous (sometimes absent). First-winter told from smaller Bay-backed Shrike by rufous-buff rump and uppertail-coverts; lacks chestnut on back of latter. **Voice** Usually silent. **Habitat** Open scrubland and cultivated regions. **Note** Rare passage migrant and winter visitor to Oman; vagrant Kuwait, UAE.

PLATE 75: ORIOLES, DRONGOS AND BABBLERS

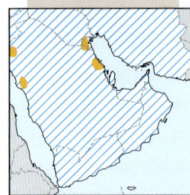

Eurasian Golden Oriole *Oriolus oriolus* PM
L: 24. Male unmistakable; female and immature male greenish above with olive-brown wings and tail, rump yellowish-green, underparts yellowish-white, indistinctly dark-streaked. Appears relatively short-tailed in flight, with woodpecker-like undulations, but strong and fast; often changes direction, tilting and angling wings accordingly. Heard more often than seen. Indian Golden Oriole *O. kundoo* may visit (collected Iran; reported, but not accepted, Oman); longer-billed with black extending behind eye, male with tertials and inner secondaries broadly tipped yellow; female and immature with extensively yellowish-washed underparts and finer streaks than Eurasian. **Voice** Song loud yodelling *tjoh-wlee-kleeooh*, the last note (often given alone) emphasised and descending; alarm hoarse, mewing *kra-eik*. **Habitat** Parks, gardens, broad-leaved woodland. **Note** Passage hatched; occasional in winter Oman, UAE.

Black-naped Oriole *Oriolus chinensis* V
L: 26. Similar to Eurasian Golden Oriole, *but black line through eye and nape in both male and female*. Male is bright yellow with greenish-yellow back and wing-coverts. Female is similar but greenish-yellow instead of yellow. **Voice** Harsh, cat-like call. **Habitat** Parks, gardens, broad-leaved woodlands. **Note** Vagrant Oman.

Ashy Drongo *Dicrurus leucophaeus* V
L: 29. Similar to Black Drongo but plumage greyer and *bill proportionally longer and slimmer; lacks white rictal spot*. Adult glossy dark slaty-grey above; dark grey below; ear-coverts contrasting darker depending on light. First-winter birds, palest on belly, have matt smoky-grey underparts *without white fringes*, latter also absent above (white blotches or fringes present above and below in first-winter Black Drongo); undertail-coverts white-tipped, eye fairly bright red (garnet in young Black Drongo). Crown flattish, nape slightly peaked (more rounded in Black Drongo). *Hunts from within or under canopy*, less often from exposed open perch. Migrant subspecies *longicaudatus* recorded. **Voice** Alarm an abrupt, short, harsh, dry rattle. **Habitat** Woodland, parkland. **Note** Vagrant Kuwait, Oman, UAE.

Black Drongo *Dicrurus macrocercus* V
L: 28–30. Adult glossy blue-black; long, deeply forked tail curves outwards towards end; *small white rictal spot in all ages, bill heavy*. First-winter duller with white tips on mantle, flanks, rump and especially belly feathers. Sits upright on conspicuous perch, making aerobatic aerial sallies in pursuit of insects. **Voice** Harsh, throaty *schweep-schweep*, also high-pitched whistles and clicks; vagrants usually silent. **Habitat** Trees, scrub, cultivation, often near habitation and water. **Note** Formerly bred SE Iran; vagrant Iran, Oman, UAE.

Arabian Babbler *Turdoides squamiceps* RB
L: 26. Greyish-brown, lightly streaked, often with head appearing 'moth-eaten'; faint dark mottling on throat and breast; bill blackish with paler base, legs brownish to dark grey. (Birds in Yemen lowlands and some highland areas have variable off-white face with whitish eye-surround and bill varying from orange-red to yellowish-orange.) Most often in close-knit groups, on ground, in bush or tree. **Voice** Typical squeaky or piping calls, including high-pitched *piu-piu-piu-piu-piu*, decelerating towards end. **Habitat** Dry scrubby areas, wadis, arid hills and open wooded savanna, especially acacia, from sea level to 2,400m; also irrigated plantations, shelterbelts. **Note** Recent range expansion in Arabian Peninsula.

PLATE 76: PARADISE FLYCATCHER, HYPOCOLIUS, PENDULINE TIT AND BULBULS

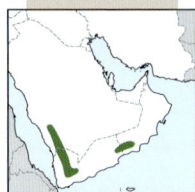

African Paradise Flycatcher *Terpsiphone viridis* RB
L: Male 30–36 (includes long tail), female 20. Unmistakable, bulbul-sized *with chestnut upperparts, glossy blue-black head and breast*, with rest of underparts slaty; *often large white wing-patches*. *Male has very long central tail feathers*, absent in less richly coloured female. White morph, rare in Arabia, has white back, rump, tail and wing-coverts. When flycatching, flight slow, heavy and wavering; tries to be secretive. **Voice** Harsh *scheep*; hoarse *tseaeae-tsceaeat*. Song has Eurasian Blackbird-like quality, *twe, twoo, twoo, twoo, twoo* uttered with fanned tail jerking from side to side **Habitat** Semi-tropical woodlands, 200–2400m; usually near water. Nests in fork of tree. **Note** Resident in S Oman.

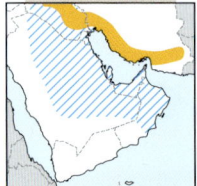

Grey Hypocolius *Hypocolius ampelinus* WV
L: 23. Recalls slim Southern Grey Shrike but with longer tail. *Sleek, soft blue-grey male, with black-tipped tail and black eye-mask joining over nape* (obscure in young male); primaries black (barely showing at rest) with *pure white tips, prominent in flight*. Female and immature featureless grey-brown, lacking black on head and with only diffuse dark tip to tail. May raise slight crest. Can be tame; often remains still, lurking in cover. Will fly high with steep climb, often for a distance, and flocks will circle for several minutes; note long tail, short wings and rather rapid wingbeats. **Voice** Mellow flight call, a liquid *tre-tur-tur*, notes running together and the last two lower pitched. When perched, a descending *whee-oo*, like Eurasian Wigeon. **Habitat** Fruit-bearing trees, scrub, palm groves on passage and in winter. **Note** Winter hatched (some birds resident in breeding range); rare Oman.

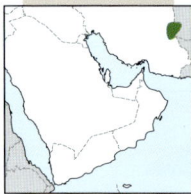

Black-headed Penduline Tit *Remiz macronyx* V
L: 11. Adults of two races, *neglectus* breeding in south Caspian and darker *nigricans* breeding in the Sistan region of Iran, differ from Eurasian Penduline Tit (not yet recorded in Oman) in having an *all-blackish head, the black extending onto the throat*, with a white collar in *neglectus* (lacking in *nigricans*, which is entirely dusky-buff below). Immature male and juvenile have a greyish head. **Voice** More often heard than seen. A rather quiet, piping, drawn-out and falling *seeeeee*. **Habitat** Migrants occur in trees, bushes or reeds near water. **Note** Vagrant Oman.

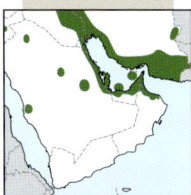

White-eared Bulbul *Pycnonotus leucotis* RB (E/I)
L: 18. Easily told by black head and throat, and *large white cheek-patch*; often shows slight crest; undertail-coverts yellow, longish tail with noticeable white tip. Juvenile has browner head than adult. Rump can be pale, as in Red-vented Bulbul, with which it hybridises readily, the progeny having variable intermediate characteristics including indistinct cheek-patch, dark spotting on upper breast and orange undertail-coverts. Often makes flycatching sallies; gregarious. **Voice** Lively bubbling jumble of notes, with simple repetitive tune; more musical than Red-vented, *too-tiddly-ooo; twee-ooo-wee-ooo* and longer bubbling variations *doo-widdly-iddly-wick*. **Habitat** Woodland, parks and gardens, urban and rural settings near human habitation. **Note** Native, but now widely introduced in Arabia. [Formerly White-cheeked Bulbul *P. leucogenys*]

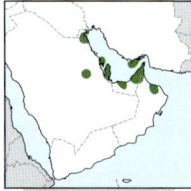

Red-vented Bulbul *Pycnonotus cafer* RB (E/I)
L: 22.5. Sooty-brown with black head, slight crest, and *fine pale scalloping on upperparts and breast; undertail-coverts red*. In flight, reveals *off-white rump* (White-eared Bulbul's rump rarely as pale) and white tip to long blackish tail. Often makes flycatching sallies; gregarious. **Voice** Noisy; calls bubbly or a burbling chatter and fairly loud, including typical *pick-yow-you* or *pee-who*. **Habitat** As White-eared Bulbul, but intolerant of drier native wooded habitats. **Note** Not native; breeding populations originate from escapes.

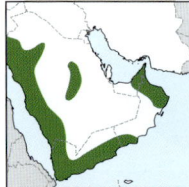

White-spectacled Bulbul *Pycnonotus xanthopygos* RB
L: 19. Size of small slim thrush, often noisy, with rather floppy flight action. Drab with *sooty-black head* shading into grey-brown upperparts and paler greyish underparts, with an *obvious white eye-ring*. Tail rather long, dark brown *without white tip; undertail-coverts yellow*. Crown feathers often slightly raised. Sociable, can occur in large groups. **Voice** Fluty and fitful, but rather loud and obvious *bli-bli-bli-bli* or *bul-bul-bul-bul-bul*. Calls include a loud, rather harsh *pwitch* and *trratsh*. **Habitat** Fruiting trees, gardens, palm groves, wadis with cover. **Note** Locally common native. [Formerly Yellow-vented Bulbul]

PLATE 77: MAGPIE, CROW AND RAVENS

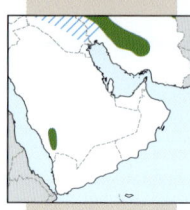

Eurasian Magpie *Pica pica* — V, E/I
L: 48. Unmistakable; *glossy black and white with long, graduated glossy-green tail; in flight, shows short, rounded wings with white primaries bordered black*. Flight action energetic with irregular flapping wingbeats. Often congregates in flocks, with so-called 'parliament' on ground. **Voice** Calls include chattering alarm *chack-chack-chack-chack-*, weaker *ch-chack* and squealing *keee-uck*. **Habitat** Bushy or open country with tall trees, scattered woods, pine and juniper-covered slopes up to 2900m. **Note** Winter dispersal hatched; vagrant Oman.

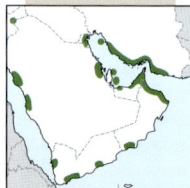

House Crow *Corvus splendens* — RB
L: 43. Readily told by *steep forehead, domed crown, grey nape, neck and breast clearly demarcated from black face*, but merging with rest of black plumage. Some individuals (adults of different subspecies, apparently including introduced/stowaway Sinhalese subspecies *protegatus*, and some, probably young birds) show little grey on head or breast. Most Arabian populations (subspecies *zugmayeri*) consistently show much grey. Wings broad, tail long and with relatively light and nimble flight with slow or fast wingbeats. Bold, noisy and gregarious, often in very large numbers, especially at roosts. **Voice** High- or low-pitched; harsh *grehr* or *waaa waaa waaa*; often also higher *aah-aah*, recalling archetypal crow heard in spaghetti Westerns. **Habitat** Ports, coastal towns and villages. Nests semi-colonially in trees or manmade structures. **Note** A pest species in the region [Alt: Indian House Crow]

Brown-necked Raven *Corvus ruficollis* — RB
L: 50. Larger than Fan-tailed Raven with *proportionately longer, slimmer wings, longer head* and *slimmer bill* (often held drooping in flight), and *bronzy-brown sheen on nape and neck* (can be difficult to see); longish, wedge-shaped tail often shows central feathers protruding beyond tail outline. *At rest wings reach to or beyond tail-tip*. Juvenile lacks brown on neck. In pairs or flocks, roosts communally post-breeding. **Voice** Croaking *raark*. **Habitat** Deserts, semi-deserts, arid mountains, often near remote habitation, camps, livestock enclosures, villages. **Note** Absent from southernmost Oman (Dhofar mountains) where replaced by Fan-tailed Raven.

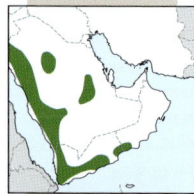

Fan-tailed Raven *Corvus rhipidurus* — RB
L: 47. Readily told from other crows (though only Brown-necked Raven and House Crow occur in the same range) by *very short tail and bulging trailing edge, giving it an unmistakable, almost 'flying backwards' (vulturine) flight silhouette*. Strong bill is shorter and heavier than Brown-necked Raven. When soaring overhead black coverts contrast with slightly paler flight feathers and the greyish feet may be visible. *At rest wings extend well beyond tail-tip*. Often congregates in large groups, which will soar, raptor-like, in thermals or updrafts. **Voice** High-pitched, rather gull-like croak. **Habitat** A wide variety of habitats from sea level to over 3000m; often close to human habitation. Nests on ledge or hole in rock face. **Note** Some winter dispersal; replaces Brown-necked Raven in S Oman; vagrant UAE.

PLATE 78: LARKS I

Greater Hoopoe-Lark *Alaemon alaudipes*　　　　　　　　　　　　　　　　RB
L: 18.5. Large, slender sandy-buff lark, with *long tail, long decurved bill and variable black band across mostly white inner wing*. Underparts whitish, often spotted black on breast. Juvenile with few breast-spots and shorter, less decurved bill. Runs speedily with sudden stops in upright position; also creeps away slowly. In characteristic songflight, male ascends vertically a few metres, twists over and spirals to bush or ground with outstretched wings. Solitary or in pairs, not mixing with other larks. **Voice** Song, mostly at sunrise, melodious and melancholy, starting slowly, accelerating and ascending, then dropping and dying away slowly *dee-dee-dee-dee-dee, dee, de-de-de-de-de-dee-dee*; also distinctive *weerrp* or *jeerrp* on territory. **Habitat** Sandy desert, semi-desert, coastal dunes. **Note** Some post-breeding dispersal.

Thick-billed Lark *Ramphocoris clotbey*　　　　　　　　　　　　　　　　V
L: 17. *Unmistakable, sandy-grey lark with large head, enormous, swollen pale bill, and bold black-and-white patterned sides of face and neck; underparts whitish with bold black spots or streaks*. Stance upright; may run at high speed. In flight, *long wings show blackish underwing with broad white band along trailing edge; short tail has dark band near tip*. Juvenile lacks black on head, neck and underparts which are creamy-white; upperparts pinkish-grey; wing pattern as adult. In low, undulating flight appears long-winged with large rounded head. **Voice** Song, from ground and in flight, jingling with some quiet, sweet warbling notes. Flight calls include a quiet *peep* or *co-ep*, a plaintive *swee* and a soft *blit-blit*; on landing *shrrreeep*; on ground *woot-w-toot*. **Habitat** Stony deserts; post-breeding also grassy wadis, rocky slopes and cultivation edge. **Note** Some dispersal in winter; vagrant Oman, Yemen.

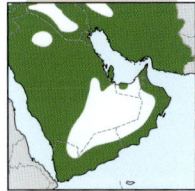

Desert Lark *Ammomanes deserti*　　　　　　　　　　　　　　　　　　　RB
L: 15. Short-tailed with *broad, rounded wings and slow, floppy, undulating flight. Unstreaked above* except for vague mottling on mantle in some; underparts buffish or greyish-white, sometimes unmarked, but often diffusely streaked on breast; rufous flight feathers and underwing in some birds. *Tertials fall clearly short of wing-tip. Longish, stout pointed bill*, yellowish-horn with dark culmen, gently tapering. There is considerable variation in this species, often in response to habitat (darker birds on black lava soils and paler birds on sandy substrates); many subspecies have been described. Races in Oman not fully determined, but likely to include *taimuri* (Hajar mountains), *samharensis* and perhaps *saturata*. The extralimital races illustrated show a range of typical plumages for the species. **Voice** Typically short, soft and melodious *dee-leeut*. Song includes phrases of call, given from ground or in descending glide. **Habitat** Arid hills, stony or rocky slopes with sparse vegetation. **Note** Some post-breeding dispersal.

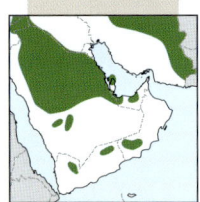

Bar-tailed Lark *Ammomanes cinctura*　　　　　　　　　　　　　　　　　rb
L: 13. Resembles Desert Lark but smaller, with more upright stance, rounder head, *smaller bill, clear-cut black band to tip of reddish-brown tail*, and rufous-buff *wings having blackish tips*. In subspecies *arenicola* (Near East/Arabia) unstreaked upperparts are pale sandy-rufous. Runs faster than Desert Lark with more abrupt stops. From Dunn's Lark *by much smaller bill, different tail pattern, unstreaked upperparts and crown*, and blackish wing-tip projecting beyond tertials; also habitat. **Voice** Song can be mistaken for Black-crowned Sparrow-Lark, like swinging pub sign creaking in wind, 2–3 notes *tlee-tloo-hee*; also *dee-dee-doo*. Flight call purring, soft hoarse *twer*; also *see-oo*. **Habitat** Flat or undulating desert with scattered vegetation and gravelly or stony rises. **Note** Some post-breeding dispersal.

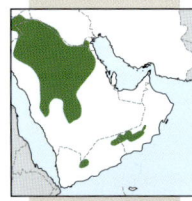

Dunn's Lark *Eremalauda dunni*　　　　　　　　　　　　　　　　　　　　rb
L: 14. Small, sandy rufous-brown with dark streaks on crown and, vaguely, on mantle; black sides to sandy-brown tail (white sides in Greater and Lesser Short-toed Larks), and large pinkish bill with pronounced curve near tip. Long tertials almost reach wing-tip. Broad whitish eye-ring bordered below by dark line, with a dark moustache and line behind eye. Broad, rounded wings and relatively short tail in flight. Appears large-headed, with upright stance; runs fast for short distances with sudden stops. **Voice** Flight call drawn-out, soft *wazz* or *ziup*; also thin liquid *prrrp*. **Habitat** Flat sandy or stony semi-desert with low scrub and grass. Often difficult to find. **Note** Vagrant UAE. Middle Eastern populations are sometimes split from African birds as a separate species: Arabian Lark *E. eremodites*.

PLATE 79: LARKS II

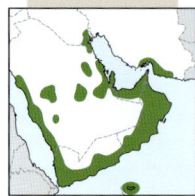

Black-crowned Sparrow-Lark *Eremopterix nigriceps* — RB
L: 12. Small stocky lark with *stout, deep-based, pale blue-grey bill with curved culmen*. Male unmistakable. Female has *unstreaked pale sandy-grey upperparts*, faintly streaked crown, pale face and hindneck, buffish underparts, faintly streaked across upper breast, and *darkish underwing-coverts*. From Bar-tailed Lark by heavier bill and unmarked tail-tip. Flight bouncing; feeds on bare ground. **Voice** Song, often uttered in circling display flight, a repetition of 2–4 loud, sweet notes *chee-dee-vee* or *pooo, pee-voo-pee*. Flight calls bubbling twitter, dry *rrrp* or soft *tchep*. **Habitat** Semi-desert, sandy or stony plains with low scrub, edges of cultivation, saltflats, coastal dunes. **Note** Nomadic outside breeding season.

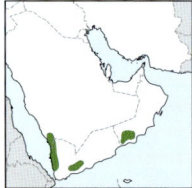

Singing Bush Lark *Mirafra cantillans* — MB, rb
L: 15. Broad-winged, fairly short-tailed lark with low, *weak, fluttering undulating flight (songflight with bat-like, jerky wingbeats)*. Yellowish-horn *bill strong, longer than deep with pronounced curved culmen. Adult has dull rufous tinge to flight feathers*, warm buff in immature. Upperparts streaked darker, buffish-white underparts with finely streaked breast, forming dark spot at breast-sides; tertials long, reaching wing-tip; white at sides of tail hard to see, as tail closed in flight. **Voice** Song from ground, post or in air, chattering but musical, ending in Corn Bunting-like jingle *ti-vit-tir-wit, che, che, che, che, che* accelerating and descending. Call a quiet *proop-proop*. **Habitat** Dry grassy and bushy plains, mountains, sparse semi-deserts; often near cultivation. **Note** Resident (or migrant breeder) in S Oman.

Eurasian Skylark *Alauda arvensis* — pm, wv
L: 18. Medium-sized with earth-grey to sandy upperparts, streaked dark with warm brown edges to tertials and coverts in fresh plumage; underparts buffish-white heavily streaked on breast; can show a small crest (much smaller than Crested Lark); head markings rather indistinct. Juvenile has upperparts spotted dark with scaly ochre markings. *In flight, shows distinct whitish trailing edge to wings* (except in juvenile) and broad triangular *tail with white sides*. **Voice** When flushed and in flight gives a variable rolled *chrriup*; or *trruwee*. **Habitat** High and low grasslands, cultivated fields; in winter more widely in open areas. Avoids deserts. **Note** Passage and winter hatched.

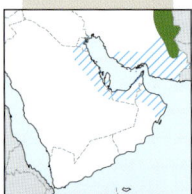

Oriental Skylark *Alauda gulgula* — pm, wv
L: 16. Similar to Eurasian Skylark, but smaller with obviously *shorter tail*, comparatively longer, more pointed bill, and very short primary projection; also slightly rusty tinge to ear-coverts and fringes of flight feathers. In flight, size of Woodlark with similarly short tail; *broad wings have inconspicuous buffish trailing edge* (white in Eurasian Skylark). Also from Eurasian Skylark by call, duller head marks, and tail having buffish-white outer feathers (bleaching to white). **Voice** Distinctive flight calls are soft *pyhp* or *twip* recalling Ortolan Bunting, and a characteristic hard buzzing *bzzeebz* or *baz-baz*. **Habitat** Open grassy and cultivated lowlands, but also grassy hills. **Note** Passage and winter hatched, but rare; probably overlooked. [Alt: Small Skylark]

Crested Lark *Galerida cristata* — RB
L: 17. Rather stocky, short-tailed lark with *long spiky crest*. Upperparts sandy-grey or rusty, diffusely streaked darker on hindneck and mantle; breast more heavily streaked. Flight flappy, *broad wings showing rusty-buff underwings; short tail blackish-brown with cinnamon sides. Lacks white trailing edge to wing*. Juvenile heavily pale-spotted above. **Voice** Clear *du-ee*, also varying fluty *ee* or *uu* sounds. Song sweet and plaintive with phrases of 4–6 repeated notes; slower, clearer and shorter than Eurasian Skylark, often includes mimicry; from exposed perch or high in air. **Habitat** Grassy or arid country, cultivated plains and semi-deserts; often near habitation, tracks and roadsides. **Note** Gathers in large post-breeding flocks.

PLATE 80: LARKS III

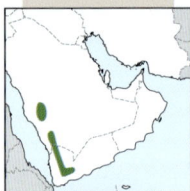

Blanford's Lark *Calandrella blanfordi*   V
L: 14. Resembles Greater Short-toed Lark; occurs SW Arabia, where Greater Short-toed usually has streaked grey crown while Blanford's is chestnut-red, usually with blackish lateral streak on forecrown, dark loral streak and dark horn-grey culmen (in Greater Short-toed, usually no dark crown-streak, lores pale and culmen generally yellowish-horn); black patches on sides of throat more horizontal, sometimes almost meeting below white throat. **Voice** Song uttered in circular, slightly undulating flight, *chew-chew-chew-chew* mixed with call-notes and fluid phrases. Call like Greater Short-toed, sometimes followed by short twitter *pit-wit-pit*; intense, whistling *peeeep*, soft *tsuru* and explosive Corn-Bunting-like *ptk* in flight. **Habitat** Open stony plateaux and fields, often with scattered bushes, mainly 1800-2500m, lower in winter. **Note** Vagrant Oman. [Formerly Red-capped Lark *C. cinerea*. Sometimes split as a separate species, Rufous-capped Lark *C. eremica*.]

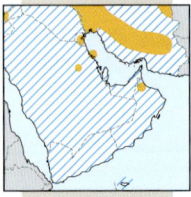

Greater Short-toed Lark *Calandrella brachydactyla*   PM,WV, mb
L: 14. Small, with streaked upperparts, buffish-white *generally unstreaked underparts with variable black patch at sides of neck* (adult); relatively stout, pointed pale bill, *long tertials almost covering wing-tip*; median coverts boldly patterned. Streaked crown greyish, tinged rufous in some males. From Lesser Short-toed Lark by *black neck-patches (in some just thinly streaked smudge), long tertials and longer bill*. Flight undulating; often in dense flocks and flies low. **Voice** Typical flight call a sparrow-like *tjirp*, and *drelit*. Song unmusical, repetitive short bursts (includes mimicry), in circling flight. **Habitat** Steppe, semi-desert, cultivated plains. **Note** Passage and winter hatched; absent from north in winter.

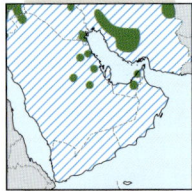

Lesser Short-toed Lark *Alaudala rufescens*   pm, wv
L: 13. Resembles Greater Short-toed Lark; but stockier, *with shorter, stubbier bill, bulging jowls, more distinctly streaked breast, lacking black patches at sides, and with wing-tip clearly exposed beyond tertials* (beware Greater Short-toed with worn tertials). Upperparts sandy-grey to rufous-brown; crown occasionally tinged rufous; supercilium generally less noticeable than in Greater Short-toed, but patterning below eye more pronounced. Flocks in winter and on passage. **Voice** Flight call abrupt, fast dry staccato *prrrrt* or *prrr-rrr-rrr*; also Eurasian Skylark-like *drrie* and quick *dreeup*. Song, in spiralling flight with unbroken deliberate wingbeats, varied, melodious and heavily mimetic. **Habitat** Steppe, saltflats, stony desert, cultivation. **Note** Passage and winter hatched.

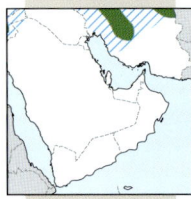

Calandra Lark *Melanocorypha calandra*   V
L: 20. Large, heavy-billed lark with relatively short tail. In flight *shows blackish underwing with conspicuous white trailing edge and white sides to tail*. On the ground *shows swollen yellowish-horn bill, black patches at sides of lower throat* (of variable shape, inconspicuous in some autumn birds) and variable whitish supercilium. Smaller Bimaculated Lark also has black neck-patches, but has paler underwing without conspicuous white trailing edge, lacks white at sides of tail, and instead has white tip. See also female Black Lark. Undulating flight low with deliberate 'wader-like' wingbeats. Flocks outside breeding season. **Voice** Flight call harsh, rolling *terrelet*; also Eurasian Skylark-like note. **Habitat** Open cultivated plains, grass and cereal fields, steppe and wastelands. **Note** Partial migrant, summer visitor to some breeding areas; winter hatched; vagrant Bahrain, Kuwait, Oman, Saudi Arabia, UAE.

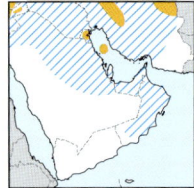

Bimaculated Lark *Melanocorypha bimaculata*   WV, PM
L: 16. Resembles small, shorter-tailed Calandra Lark; in flight has *dull grey-brown underwing without clear white trailing edge, and white-tipped tail* with outer feathers buff-brown. On ground, *upperparts show more prominent scaling* with head pattern more contrasting than Calandra, with rusty cheeks, pronounced *long white supercilium and dark lores* giving capped appearance. **Voice** Flight call recalls Calandra, *trrelit*, with rather gravelly scrunching; also a Short-toed Lark-like *dre-lit*. Song includes drawn-out rolling call-note, delivered from ground or air. **Habitat** Thinly vegetated hills or marginal stony cultivation up to 2,400m; in winter down to sea level in agricultural areas, often near penned livestock camps. **Note** Partial migrant; passage hatched; some winter in Arabia.

PLATE 81: MARTINS

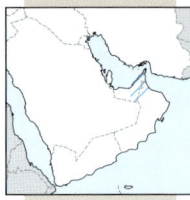

Grey-throated Martin *Riparia chinensis*   V

L: 12. Small, *mouse-brown above* with greyish-white underparts *merging into darker greyish-brown upper breast, throat and head*; lacks breast-band of Sand Martin, has shorter tail accentuating broader-based appearance to wings, thus recalling larger Pale Crag Martin; differs from that species in *more fluttering flight, lack of white tail-spots and by white vent/undertail*. See also Pale Martin. Flight and behaviour similar to Sand Martin. **Voice** Soft *dree-dree* or *dree-err, dree-err*, not as dry or buzzy as Sand Martin. **Habitat** Rivers, lakes, grasslands. **Note** Rare in UAE as mapped; vagrant Oman, Saudi Arabia, Yemen. [Alt: Formerly Brown-throated Martin *Riparia paludicola*]

Sand Martin *Riparia riparia*   PM, wv

L: 12. Small, dull *brown-and-white martin* recalling Grey-throated Martin; distinguished by white underparts with well-marked *brown breast-band* (though paler and indistinct in juveniles) and slightly longer and more deeply forked tail. From Pale Crag Martin by fluttering flight action, breast-band, white lower underparts and lack of white tail-spots. Juvenile has scaly, buffier upperparts and poorly defined breast-band (see also Pale Martin). **Voice** Usual call in flight is a vowel-less, rasping repeated *tschr* or *zrrrr-zrrrr*, persistent and slightly buzzy in groups; louder *chirr* given in alarm. **Habitat** Open country with wetlands. Widespread on migration, though usually over water or grasslands. Nests colonially in tunnel excavated in sandy bank. **Note** Passage hatched; some winter in S Arabia.

Pale Martin *Riparia diluta*   V

L: 11.5. Marginally smaller than Sand Martin (with which may consort, although respective flocks typically remain separate), being pale *mousey grey-brown above, with breast-band ill-defined or absent*, and *less deeply forked, almost notched, tail*. Also has slightly paler underwing; off-white throat, with *ear-coverts smudged dingy-brown and their lower edge not clear-cut*. Note young Sand Martin often has poorly defined breast-band, and some individuals perhaps not safely separable. **Voice** Usually silent in winter but quiet, sweet chatter sometimes audible; not heard to give the 'buzzing' characteristic of Sand Martin. **Habitat** As Sand Martin, over water or irrigated grasslands. **Note** Passage and winter hatched (variable numbers annually); vagrant Oman.

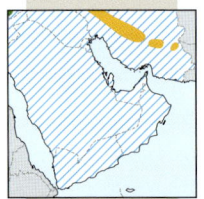

Eurasian Crag Martin *Ptyonoprogne rupestris*   pm, wv

L: 14.5. Large broad-winged, brown martin, almost square-tailed, similar to smaller and typically paler Pale Crag Martin but l*arger, more heavily-built, with darker upperparts lacking greyish tinge to rump; cheeks fairly dark, usually contrasting with pale throat and underparts; fine spots on chin and throat* (can be hard to see). Underparts buffish-grey becoming darker towards undertail; white spots in spread tail as in Pale Crag Martin but *underwing-coverts distinctly darker than flight feathers*. Flight more swooping and diving than in most other hirundines. **Voice** Song a quiet twittering; in flight a short *chip* or *chirr*. **Habitat** Mountain gorges and rocky inland and coastal cliffs. On passage in more open country. **Note** Passage hatched, but rare in south; some winter in Arabia.

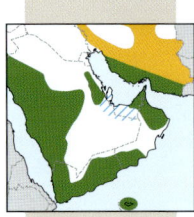

Pale Crag Martin *Ptyonoprogne obsoleta*   RB

L: 12.5. Similar to larger Eurasian Crag Martin but usually discernibly *smaller and paler*, more grey-brown; upperparts, especially *back and rump, appear slightly greyer than wings; underparts off-white including chin* (lacking dark spots of Eurasian Crag Martin) merging into pale mouse-grey undertail-coverts; *less contrasting head pattern* though ear-coverts sometimes darker than crown; underwing pale grey with *brownish-grey coverts contrasting much less than in Eurasian Crag Martin* (overhead, against light, may look confusingly close to Eurasian Crag). **Voice** Dry, rather quiet twittering and repeated single *drrrrt*. **Habitat** Hilly country with gorges and ravines, but often also lowlands near habitation in sand desert, towns and cities, where it will nest on low or high-rise buildings. Often over wetlands near breeding areas. Builds open half-cup-shaped nest in caves, on cliff or under eaves of building. **Note** Vagrant Kuwait. [Alt: African Rock Martin]

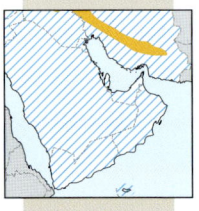

Common House Martin *Delichon urbicum*   pm

L: 12.5. Smaller than Barn Swallow; easily recognised by *bluish-black upperparts with striking white rump and all-white underparts*; short, forked tail with white undertail-coverts and rather dark underwings. Juvenile duller with brownish wash on head and breast-sides. Flight more fluttering than Barn Swallow with long glides often high in air. Frequently perches on wires. Gregarious, migrating in flocks. **Voice** Commonest flight call a short dry warbling *prrlit*, often sounding conversational. Gives repeated high trilled *scheeer* in alarm. **Habitat** Mountains, hills, villages. Mud nest built under eaves of house, on cliff or in cave. **Note** Passage hatched; some winter in Arabia.

PLATE 82: SWALLOWS

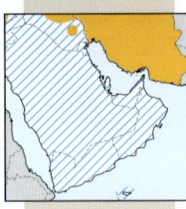

Barn Swallow *Hirundo rustica* — PM, WV

L: 16 (including tail streamers). Easily distinguished by *bluish-black upperparts, long tail-streamers,* (small *white patches visible in tail* when spread), *chestnut forehead and throat, and solid dark breast-band*; underparts buffish-white including underwing-coverts. Some subspecies have underparts reddish-buff. Juvenile lacks tail streamers, has brownish breast-band and pale rusty forehead. Flight is strong and elegant with much banking and turning, often hunting rather low. Often perches on wires with other hirundines. **Voice** Song a melodious twittering; contact call *witt-witt*, alarm call disyllabic *tsi-wit*. **Habitat** Open cultivated country with settlements; over any area on migration. Nests on ledge in building. **Note** Passage and winter hatched, but absent from north in winter.

Wire-tailed Swallow *Hirundo smithii* — pm, wv

L: 16 (including tail streamers). Slightly smaller than Barn Swallow, without breast-band. Clean-coloured with rich *blue upperparts and eye-mask*, white tail patches as in Barn Swallow, *rufous-chestnut crown* and rather square-ended tail with long, fine outer tail feathers (shorter in female and often difficult to see in flight); *underparts gleaming white including underwing-coverts* with bluish flanks and broken vent-bar. Juvenile Wire-tailed Swallow lacks tail streamers, has almost square tail, dull bluish wings, brown back and crown with darker eye-mask; underparts as adult. Flight similar to Barn Swallow, but with *more compact, triangular outline* discernible. Often joins other hirundines. **Voice** Call a sharp *tchik*. **Habitat** Along rivers, over lakes, grasslands. **Note** Winter hatched, but rare.

Lesser Striped Swallow *Cecropis abyssinica* — V

L: 19. Rather slim with deeply-forked tail. Only confusable in region with Red-rumped Swallow but easily told by *chestnut crown and rump, black streaks on* underparts (variable, but usually only faintly streaked on breast in some Red-rumped Swallow), *pale chestnut underwing-coverts and a broad white band on undertail.* Juvenile has shorter outer tail feathers. **Voice** Descending series of squeaky notes. **Habitat** Open areas, as other swallows. **Note** Vagrant Oman (from Africa).

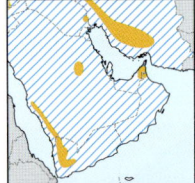

Red-rumped Swallow *Cecropis daurica* — pm, wv

L: 17 (including tail streamers). Closely resembles Barn Swallow but easily told by *rufous collar and broad pale rufous rump*; underparts, including underwing-coverts, buffish-white, faintly streaked (but variably so, streaks obvious in some, almost absent in others), *undertail-coverts striking black; lacks breast-band and white in tail.* Juvenile browner with paler collar and rump, and shorter outer tail feathers. Flight slow and graceful, frequently gliding for longer periods than Barn Swallow, with tail flared or closed. **Voice** Usually rather quiet. Song shorter and quieter than Barn Swallow; in flight a short soft nasal *tweit*, reminiscent of House Sparrow. **Habitat** Inland and sea cliffs, cultivated areas; in flat country frequents bridges and buildings. *Builds flask-shaped mud nest with spout-shaped entrance* in caves, under overhang, bridge or on building. **Note** Has bred UAE; passage hatched; some winter in S Arabia.

Streak-throated Swallow *Petrochelidon fluvicola* — V

L: 11. Diminutive and front-heavy; dusky throat and upper breast seen at close quarters is strongly dark-streaked; thighs with dark flecks. Crown dingy brown (immature) to ginger-chestnut (adult), rump contrastingly brownish. White streaks on mantle in adult. Tail only slightly forked. Viewed overhead can resemble Brown-throated Martin, but more compact and thickset. Often joins flocks of other hirundines. **Voice** Not heard from vagrant individuals. **Habitat** Lakes, ponds and open country. **Note** Vagrant Oman, UAE (from Asia). [Alt: Indian Cliff Swallow]

PLATE 83: MISCELLANEOUS WARBLERS

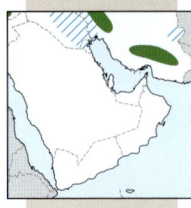

Cetti's Warbler *Cettia cetti*   V
L: 13.5. Short-winged, *round-tailed*, rather robust-looking, *warm brown warbler*; keeps well-concealed in thickets and more often heard than seen; *song distinctive*. Off-white below, tinged rufous on flanks and breast-sides; *vent and undertail-coverts brown with feathers fringed white*; uniform chestnut-brown tail often jerked or held cocked; rather thin but *distinct white supercilium*. **Voice** Abruptly delivered song from well-concealed perch is a sudden, loud, explosive outburst *plit... plitiplitipliti... plhi-(pliti)*. Alarm call hard, forceful *tlitt*, or rapid, rattle, *tlitt-tlitt-tlitt*. **Habitat** Dense vegetation, bushy thickets, mostly near streams, ditches and reedbeds. **Note** Winter dispersal hatched; vagrant Oman, UAE.

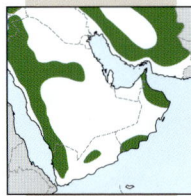

Streaked Scrub Warbler *Scotocerca inquieta*   RB
L: 11. Recalls Graceful Prinia. Furtive; often in pairs or small groups and on the ground. *Long, scarcely graduated tail cocked and constantly manoeuvred*; differs from Graceful Prinia in having blackish tail, tipped white on underside of outer feathers. *Distinct white supercilium, dark eye-stripe* and flat crown; *breast finely streaked*; bill yellow-horn and legs rather long. **Voice** Song thin *di-di-di-di-di*, descending *di-di-di-de-de*, also dry *dzit, dzit* followed by warbling *toodle toodle toodle*. Calls include *drzip, dri-dirrirri*, loud rolling *tlyip-tlyip-tlyip*, sometimes fast and descending; also scolding *prrt*; alarm-note, sharp piping *pip*, often repeated. Contact call characteristic, disyllabic, clear *dee-düü*, latter note lower. **Habitat** Rather barren, stony hillsides, semi-deserts and sandy plains with low scrub, up to 2,600m. Builds domed nest in low bush. **Note** Descends to lower elevations in winter. [Alt: Scrub Warbler]

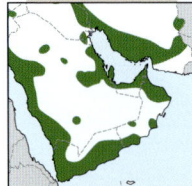

Graceful Prinia *Prinia gracilis*   RB
L: 11. Small, active, variable (across subspecies) grey-buff warbler with *long graduated tail*, frequently cocked and slightly fanned; narrowly streaked on crown, mantle and back; roundish head and *pale face with prominent eye, but without supercilium and eye-stripe* (unlike Scrub Warbler); tail dark above, paler below with *black-and-white tips*; underparts off-white, unstreaked (streaked in Scrub Warbler); bill fine, black in breeding male, brown in female. Skulks in cover, but usually confiding. Flight weak, the long tail being obvious. **Voice** Rather noisy. Song monotonous, winding *drrir-drrir-drrir-drrir-drrir* or fast *di-der di-der di-der di-der*; chittering calls include an abrupt *chlip; churr-churr*, metallic winding *srrrrrrt*, and hard *tsiit* or *chig* in alarm. **Habitat** Scrub and low vegetation; cultivated areas, gardens, waste ground in villages and towns. Builds domed or cupped nest in low bush or thick grass. **Note** Absent from much of central Oman.

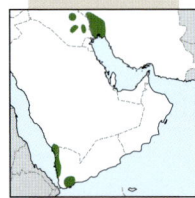

Zitting Cisticola *Cisticola juncidis*   V
L: 10. Small, short-tailed warbler, *most often seen in songflight*. Crown and upperparts streaked yellow-buff and black; face pale with prominent dark eye; tail rounded, *often fanned in flight, with black-and-white tips*, most obvious from below in songflight. **Voice** Song diagnostic, in bounding circular flight; consists of *tzip* notes repeated at regular intervals of a half to one second, with one *tzip* on each bound; also a short *kwit* from cover. **Habitat** Fields, grassland, stream-sides, marshes. Purse-shaped nest suspended in bush, long grass or crop. **Note** Vagrant Kuwait, Oman. [Alt: Fan-tailed Warbler]

PLATE 84: *PHYLLOSCOPUS* WARBLERS I

Willow Warbler *Phylloscopus trochilus* — PM

L: 11. Similar to Common Chiffchaff, with *longer primary projection; darker, pale-edged flight feathers and tertials; yellow wash on throat and breast; more distinct supercilium and pale-centred ear-coverts* (*acredula* greyish above, whitish below; *yakutensis* similar, with grey-streaked breast). *Legs pale pinkish-brown*; tail-dipping much less pronounced. *First-winter shows distinctly yellow underparts.* **Voice** Song pleasant *descending whistled verse dying away at end*. Call rising, disyllabic *hoo-eet*. **Habitat** Woodland, scrub. **Note** Passage hatched.

Common Chiffchaff *Phylloscopus collybita* — PM, WV

L: 11. Several subspecies occur, often inseparable. All show *short supercilium*, broken eye-ring (conspicuous in autumn) and *dark legs; tail frequently dipped*. **Habitat** Woodland, scrub. **Note** Passage and winter hatched. At least two subspecies are separable in the field:

Scandinavian Chiffchaff *P. c. abietinus* (winters abundantly south to Arabia). Paler and greyer above than *collybita*; buff and yellow on breast reduced, undertail-coverts whiter. **Voice** Song repeated *chiff-chaff-chiff-chaff...*, varying geographically in structure and clarity. Wintering birds also call *peep*.

Siberian Chiffchaff *P. c. tristis* (many reach Oman in winter). *Greyish-brown above, lacking yellow except for bend of wing and underwing, olive-green tinge on rump, wing-coverts, scapulars and edges of flight and tail feathers. Thin supercilium often distinct, off-white to buff, rarely a yellow trace*, ear-coverts with rusty hue; entire underparts off-white, suffused buff on breast-sides and flanks; *legs and bill almost black*. In fresh plumage *may show indistinct wing-bar* on greater coverts. **Voice** Song distinctive; fast, melodious, multi-syllabic rising and falling stream; almost unrecognisable as a chiffchaff. Call near-monosyllabic, plaintive *eep*.

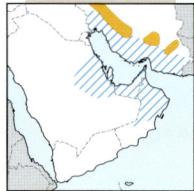

Plain Leaf Warbler *Phylloscopus neglectus* — PM, WV

L: 9. Smallest *Phylloscopus*, with short tail and lacking obvious markings or any yellow; constantly flicks wings. Plumage like greyer chiffchaffs, with brownish-grey upperparts tinged olive, shawl greyish; indistinct, short, pale buff to cream supercilium, narrow white eye-ring, dusky lores and eye-stripe, and buff-flecked ear-coverts. Underparts off-white, washed creamy on flanks. Legs and fine bill dark brown. Forages actively in cover; often hovers outside or below canopy. **Voice** Short, quiet tinkling song often heard in winter, *pt toodla toodla*. Call excitable, hard *t-jick* or *tdd*, commonly repeated. **Habitat** Scrub and trees in winter. **Note** Winter hatched; common N Oman and Musandam.

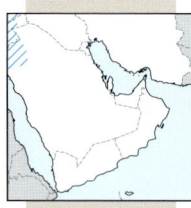

Eastern Bonelli's Warbler *Phylloscopus orientalis* — V

L: 12. Size similar to Willow Warbler but *mantle and more rounded head distinctly grey*, contrasting dark tertials (fringed yellowish) with *obvious green edges to wing and tail feathers*; underparts silky-white; *face 'washed-out' with large dark eye* accentuated by narrow white eye-ring, usually *pale lores, pale ear-coverts and indistinct supercilium*. If seen well, *yellowish-green rump distinctive*. Confusable with pale Common Chiffchaff, but has plain head and longer primary projection; told from Booted Warbler by face pattern and well-marked tertials. **Voice** Call a short, flat *chip*. **Habitat** Woodland, scrub. **Note** Passage hatched; vagrant Iraq, Kuwait, Oman, Saudi Arabia (Gulf).

Wood Warbler *Phylloscopus sibilatrix* — pm

L: 12.5. Large, *long-winged Phylloscopus*, brightly coloured, showing *vivid green upperparts with yellow supercilium, ear-coverts, throat and upper breast contrasting with gleaming white underparts; yellow-fringed secondaries form pale panel contrasting with blackish, whitish-fringed tertials*; black alula, yellow bill and pale legs. From Green Warbler by different wing pattern including lack of wing-bar; from other *Phylloscopus* also by very long wings, striking plumage, particularly head pattern and brightly fringed secondaries. Rare greyer individuals, recalling Eastern Bonelli's Warbler, told by strength of supercilium, long wings (still contrastingly fringed) and larger size. Rather active, often high in trees. **Voice** Call a clear melancholy *deeu-deeu-deeu* but usually silent on migration. **Habitat** Woodland and tall trees, also scrub. **Note** Passage hatched; scarce in Arabia.

PLATE 85: *PHYLLOSCOPUS* WARBLERS II

Dusky Warbler *Phylloscopus fuscatus*   V

L: 11. Small, *dark brown Phylloscopus* with distinct head pattern resembling brownish Common Chiffchaff. *Upperparts plain brown with rather distinct long, pale rufous-brown or whitish supercilium*, narrow and often palest in front of eye. Underparts off-white, *tinged pale rufous-buff on breast, flanks and undertail-coverts*. Fine bill with largely yellowish lower mandible; thin legs flesh-brown. From Siberian Chiffchaff by paler legs and bill. Often on ground. **Voice** Loud, hard, abrupt *chek* or *tack*, recalling Lesser Whitethroat, often persistent. **Habitat** Low cover, grass or scrub, often in moist or marshy localities. **Note** Vagrant Kuwait, Oman, Saudi Arabia, UAE, Yemen.

Yellow-browed Warbler *Phylloscopus inornatus*   pm

L: 10. Very similar to Hume's Leaf Warbler, generally brighter *with greener upperparts, whiter underparts, lacking buff suffusion; supercilium very long, pronounced and yellow-white*, rather evenly broad, *usually two distinct yellow-white wing-bars contrasting with dark-centred greater coverts*, also *stronger contrast in tertial pattern; base of bill obviously paler*. Occasionally shows faint paler crown-stripe. In worn plumage greyer with paler wing-bars, and tertial-fringes narrower and whiter. **Voice** Thin song is high-pitched *tsee tseoo-tseee*, lacking buzzing quality of Hume's; call high-pitched lisping *tsweest* or *weest* with *distinct upwards inflection*, frequently almost disyllabic *wii-ist*. **Habitat** Trees, bushes. **Note** Rare passage and winter as hatched; vagrant Iran, Qatar, UAE.

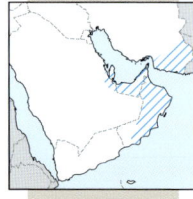

Hume's Leaf Warbler *Phylloscopus humei*   pm, wv

L: 10. Confusable with Yellow-browed Warbler; *generally greyer, duller, less yellow-green* with different call and song. In fresh plumage, upperparts and crown greyish-olive with very long, pronounced, *buffish-white supercilium*, often broadest over and in front of eye; two wing-bars, lower one distinct, *upper one short and ill-defined or absent, underparts dusky-white, tinged buffish on ear-coverts, lower throat and sides of neck; bill dark* (but pale base from directly below). Occasionally shows faint paler crown-stripe. In winter/spring worn plumage becomes greyer, wing-bars and tertial edges narrower, with upperwing-bar often lacking. **Voice** Calls sweet *wesoo* or shorter *dweed*, both *lacking upwards inflection*; in winter more disyllabic, clean *sooit*. Song thin, high-pitched, drawn-out, descending buzzing *tzeeeee*. **Habitat** Trees, bushes. **Note** Winter hatched, but rare; vagrant Bahrain, Iraq, Kuwait, Qatar, Saudi Arabia.

Arctic Warbler *Phylloscopus borealis*   V

L: 12. Rather large, *dark greyish-green Phylloscopus* with strong pale bill, striking head pattern with distinct *long narrow supercilium from just in front of eye to well behind dark-mottled ear-coverts* (sometimes kinked upwards at end), accentuated by *distinct dark eye-stripe from base of bill; short wing-bar* (sometimes two visible in fresh plumage), plain tertials, dusky flanks and pale brown legs. Confusable with smaller Greenish and Green Warblers in worn, greyer plumage; distinguished from both by size, heavier build, head pattern, plumage tone and pale legs. **Voice** Characteristic short *dzik* (or *dzi-zik*). **Habitat** Bushes, trees. **Note** Vagrant Oman, Saudi Arabia.

Green Warbler *Phylloscopus nitidus*   pm, wv

L: 11. Similar to Greenish Warbler with round head (like Common Chiffchaff), *broad supercilium from base of pale bill to behind ear-coverts and short narrow wing-bar on greater coverts* (Chiffchaffs may show inconspicuous but rather long wing-bar). From Greenish Warbler by greener upperparts, and *sulphur-yellow throat, upper breast, supercilium and ear-coverts* contrasting with dark eye-stripe. In worn plumage more greyish, almost lacking yellow tone, and wing-bar may be lost; then very difficult to tell from Greenish Warbler, but from most other *Phylloscopus* by striking head pattern, pale bill and voice. **Voice** Song loud and variable, ending abruptly, *che-wee che-wee chui chi-di chi-dit*. Call is White Wagtail-like *che-wee*, similar to Greenish Warbler. **Habitat** Bushes, trees. **Note** Passage hatched, but rare; vagrant Bahrain, Saudi Arabia, UAE.

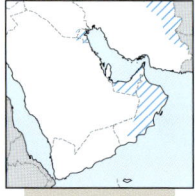

Greenish Warbler *Phylloscopus trochiloides*   V

L: 11. *Resembles a cold, greyish Willow Warbler*, with yellowish mostly restricted to *long, broad supercilia, which usually meet over bill base* (unlike Arctic Warbler) and may create capped, 'kind-faced' appearance. Also note *dull greyish-white underparts and ear-coverts*, contrasting with distinct dark eye and eye-stripe; plain tertials, *short wing-bar* (sometimes two in fresh plumage, or worn off in summer), pale bill and dark brownish legs. From Willow Warbler and Common Chiffchaff by greyish-white underparts and cheeks, stronger head markings, short pale wing-bar (if present) and yellow lower mandible (especially from Common Chiffchaff). **Voice** Call similar to Green Warbler, *chi-wee*, or shorter *chi it*. **Habitat** Bushes, trees. **Note** Vagrant Oman.

PLATE 86: *ACROCEPHALUS* WARBLERS I

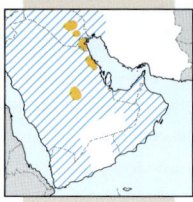

Great Reed Warbler *Acrocephalus arundinaceus* pm, wv
L: 19. Eastern subspecies *zarudnyi* occurs on passage; compared with nominate subspecies is more olive, less rufous, especially on rump, and whiter below, with a pale supercilium, dusky lores and light brown streaks on the white throat. Contrasting buffier rump shows in flight. Compared with Clamorous Reed has *longer primary projection with 7–8 primary tips showing*, proportionately shorter tail and shorter, slightly stouter bill. Moults in winter quarters. **Voice** Loud and powerful song with repetitive character; common phrase *trr-trr, karra-karra-karra, kreee-kreee-kreee*. **Habitat** Reedbeds; on passage also in drier habitats. **Note** Passage hatched.

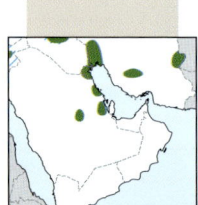

Clamorous Reed Warbler *Acrocephalus stentoreus* RB, pm, wv
L: 18. The subspecies *brunnescens* (Indian Reed Warbler) occurs. Greyish-olive in fresh plumage, whitish below with buff flanks, and long tapering pointed bill. Primary projection shorter than Great Reed. Moults post-breeding. **Voice** Loud, *strident song*, similar in tempo but more melodious than Great Reed Warbler, *witch-a-witch-a witch, chew-chew-chew-chew, skatchy, skatchy, skatchy, vachoo vachoo vachoo*, frequently including the phrase *rod-o-petch-iss*. Call a loud abrupt *tjuck* or rolled *churr*. **Habitat** Breeds in mangroves and reedbeds (including inland); also in scrub, woodland, date gardens post-breeding and in winter. **Note** Resident Iran and Arabia, dispersive or augmented by (usually biometrically larger) migrants in non-breeding season; range gradually expanding north and west.

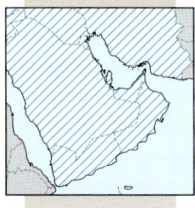

Moustached Warbler *Acrocephalus melanopogon* V
L: 13. Similar to Sedge Warbler but has broad, *clear-cut white supercilium, ending squarely on side of nape and dividing blackish crown and greyish ear-coverts; more uniform, rufous-brown upperparts* (less obviously streaked than in Sedge), with *whiter throat* and warmer brown flanks. Eastern subspecies *mimicus* occurs, which has shortish primary projection. Keeps low in vegetation or on ground where unobtrusive, *often cocking and flicking tail nervously* (also when singing). **Voice** Song with intermittent characteristic Common Nightingale-like *lu-lu-lu*. Call a loud *trr-trr*, soft, short *tcht*, longer *trr-trrrrr* and a hard *tack*. **Habitat** Reedbeds, swampy thickets. **Note** Passage hatched; vagrant Bahrain, Oman, UAE.

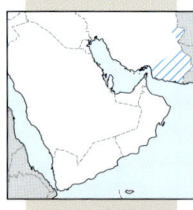

Sedge Warbler *Acrocephalus schoenobaenus* pm, wv
L: 13. Streaked, with buffish-white supercilium; confusable with Moustached Warbler and told *by buffier, less square-cut supercilium, slightly paler crown, paler ear-coverts and buffier, more streaked upperparts, which merge into warmer-coloured rump, and distinctly longer primary projection*. First-winter Sedge Warbler is yellower than adult, shows paler centre to crown and fine spotting on breast. **Voice** Song fast, often rising and falling; often starts song with rapid *trr* notes. Mimetic. Call a hard *chek* and fast, churring *trrr*. **Habitat** Reedbeds, swampy thickets; drier habitats on passage. **Note** Passage hatched.

Paddyfield Warbler *Acrocephalus agricola* pm, wv
L: 12. Slightly smaller than Caspian Reed Warbler and differing in *shorter bill, more prominent whitish-buff supercilium from bill to well behind eye* (where most conspicuous), *often with suffused darkish border above*, greyish nape shawl, dark-centred tertials with paler edges (unlike Blyth's Reed); *shorter wings, and longer tail, which when landing or on the ground is often raised slightly and constantly flicked*. Underparts show warmer wash to flanks and undertail than other unstreaked *Acrocephalus*. Bill brown with flesh-coloured lower mandible but distal part often darker in adult (so can appear dark-tipped). Crown feathers often raised; iris pale in adults. **Voice** Calls simple *chik, chik*, and a rolling *churrr*. **Habitat** Wetlands, grass or scrub. **Note** Passage hatched; vagrant Bahrain, UAE.

PLATE 87: *ACROCEPHALUS* WARBLERS II, AND *LOCUSTELLA* WARBLERS

Blyth's Reed Warbler Acrocephalus dumetorum V
L: 12.5. *Similar to Eurasian Reed and Marsh Warblers*. Identified by combination of *shorter, more rounded wings and short primary projection with six primary-tips visible* (7–8 visible in Eurasian Reed and Marsh Warblers), *uniform upperparts with little or no rump contrast, plainer tertials*, near-concolorous alula, short supercilium, which often bulges in front of eye, and dark grey bill with flesh-coloured base to lower mandible. Sometimes shows flicking and fanning movements of tail. **Voice** Song musical and highly imitative; slower, more hesitant than Marsh Warbler; often has high-pitched *lo-ly-lia* and utters *tjeck-tjeck* between phrases. Contact call a soft *thik* or *chck*. **Habitat** Bushy vegetation. **Note** Passage hatched; vagrant Bahrain, Kuwait, Oman, Saudi Arabia.

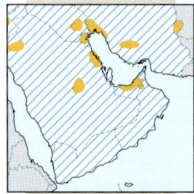

Eurasian Reed Warbler Acrocephalus scirpaceus PM, wv
L: 13. Subspecies *fuscus* (Caspian Reed Warbler) occurs. Very similar to Marsh Warbler, with similarly long wing-projection but less obviously pale edges to tertials, with *longer bill and darker legs*; greyer, less rufous above and whiter below, thus even more similar to Blyth's Reed and Marsh Warblers. **Voice** Monotonous and fairly even-pitched mixture of scratchy, grating and churring notes; *resembles song of Sedge Warbler but slower and lacks the changes in pitch and tempo*. Calls short *tchk* or hard, rolling *chrrrur*. **Habitat** Reedbeds and waterside vegetation. **Note** Passage hatched.

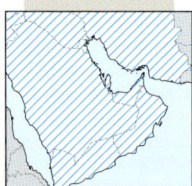

Marsh Warbler Acrocephalus palustris PM
L: 13. Difficult to identify; brown above with *slight olive tinge* and slightly warmer rump; creamy-buff below with slightly buffier flanks; short buffish white supercilium and pale eye-ring. Like all *Acrocephalus* has a rounded tail, long undertail-coverts and rather sloping forehead. *Marsh Warbler best told in spring by the pale-fringed dark tertials and long primary projection, eight primaries showing, each with pale-fringed tip*. Also buffier, less warm, than Caspian Reed (although juvenile Marsh rustier and thus extremely similar), with paler legs and shorter bill, but latter hard to judge. **Voice** *Loud, musical and full of mimicry*; in fast tempo. Calls short *chek* and distinctive short buzzy *terrrr*. **Habitat** Any cover on passage. **Note** Passage hatched.

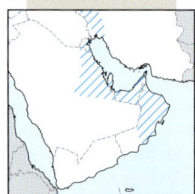

Common Grasshopper Warbler Locustella naevia pm
L: 12.5. Small, rather dark, skulking warbler with obviously rounded tail when flushed, often from underfoot. Flies to cover, in short, low, jerky flight. *Upperparts olive-brown, heavily streaked* (but appearing obscure in flight) *with faint supercilium*; underparts dirty white or yellowish with streaking on rear flanks and undertail-coverts, sometimes diffusely streaked on breast, juveniles especially. **Voice** Migrants typically silent. Sings, often at night, a dry reeling (likened to reel on fishing rod), longer, more sibilant than Savi's Warbler; infrequently gives a short *chik* call. **Habitat** Thick moist vegetation, dense bushes, thickets near streams, reedbeds; on passage, also grassland, herbaceous cover. **Note** Passage hatched, but scarce or overlooked.

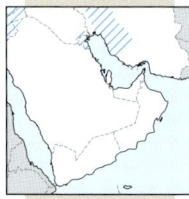

River Warbler Locustella fluviatilis V
L: 13. Secretive olive-brown warbler, with *unstreaked upperparts, diffusely streaked breast and long undertail-coverts with brownish feathers tipped white*. Confusable with Savi's Warbler but is darker and greyer, being colder, darker olive-brown above, with much more obvious darker spotting on breast (diffuse in Savi's); generally keeps low in vegetation, but will walk on ground with slightly cocked tail. **Voice** Migrants silent in region. **Habitat** Thick cover, usually near water; grasslands on passage. **Note** Passage hatched but rare; vagrant Bahrain, Iraq, Oman, Saudi Arabia, UAE, Yemen.

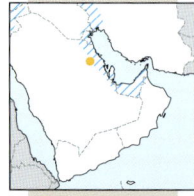

Savi's Warbler Locustella luscinioides pm, wv
L: 14. A large *Locustella*. Plumage warmer brown than River Warbler; eastern subspecies *fusca*, which occurs in region, has grey cast above, but diffusely marked on breast; also has the long undertail-coverts tipped pale, only less distinctly so than in River Warbler. Often on ground, walking slowly and stealthily with horizontal stance and jerky movements, the graduated tail raised; often bobs tail (a character shown by all *Locustella* warblers). **Voice** Song, uttered from reed stem or bush, by day or night, a monotonous, fast reeling. Call (at intruder or in alarm) also distinctive, a loud, hard, twinky pitch repeated 2–3 times. **Habitat** Reedbeds, swamps, fields, scrub and rank grass. **Note** Passage hatched.

PLATE 88: *IDUNA* AND *HIPPOLAIS* WARBLERS

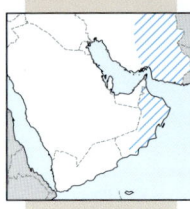

Booted Warbler *Iduna caligata*   pm, wv
L: 11.5. Small with *short, rather fine bill, rounded forehead*, recalling *Phylloscopus*, but crown slightly peaked and bill broad-based, with pale tips and edges to outer tail feathers. Tail sometimes raised quickly upwards. Lacks pale wing-panel. Confusable with Sykes's Warbler but *bill shorter, distinctly finer and usually shows dark-tipped lower mandible*; upperparts grey-brown (lacking olive); longer, *better-defined supercilium, sometimes with dark upper border, extending beyond eye*. *Legs flesh-brown* with darker feet. **Voice** Song has rhythm of Eastern Olivaceous, but faster, more melodious without quick repetition of phrases. Call recalls weak Lesser Whitethroat's *tek-tek*. **Habitat** Scrub, tamarisk and woodland, often near water; frequently in low cover, even weeds, on passage. **Note** Passage hatched, but poorly known and rare; vagrant Gulf states.

Sykes's Warbler *Iduna rama*   rb, pm?
L: 11.5. Only subtly different from Booted Warbler, having plainer face, l*onger bill, with pale lower mandible, longer tail, usually shorter primary projection (accentuating tail length), greyish-pink legs* and marginally paler underparts lacking any buff; also has more horizontal stance and *Acrocephalus*-like appearance and habits. **Voice** Song bubbly with scratchy *Sylvia* subsong-like sequences, starting with *tiju-tiju-tiju* (of tit-like character). Call hard, clicking *tak*. **Habitat** Breeds in tamarisk, damp scrub, grazed mangroves; trees or scrub on passage. **Note** Isolated breeding population in N Oman. Passage hatched, though rare; vagrant Qatar.

Thick-billed Warbler *Iduna aedon*   V
L 18. Large, plain-looking *Iduna* warbler with *short primary projection* and *long tail, somewhat resembling a shrike in shape*. Plumage warm, rusty-brown with *pale lores and no supercilium*. Bill large, rather thick, but shorter than Clamorous Reed Warbler, which it somewhat resembles, but note absence of supercilium and pale lores. **Voice** Soft *tuk*, sometimes repeated rapidly *tuk-tuk-tuk*. Also a hard *chak*. **Habitat** Woodlands, dense, thick bushes. **Note** Vagrant Oman.

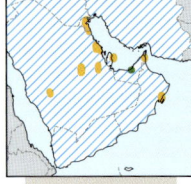

Eastern Olivaceous Warbler *Iduna pallida*   PM
L: 12.5. Recalls *Acrocephalus* warbler but greyer with square-ended or slightly rounded tail and *shorter undertail-coverts*. Upperparts *olive-brown, tinged greyish* on head and mantle, underparts buffish-white (juvenile suffused yellowish) with white throat; *wings with marked pale panel in spring at least*, tail brown inconspicuously fringed and tipped white on outer feathers. Pale eye-ring and variably distinct *supercilium from bill to rear of eye* (longer, more contrasting supercilium in Booted and Sykes's Warblers); l*ong, rather heavy, broad-based bill with yellow-pink lower mandible; legs greyish-brown*. Often deliberately flicks or *pumps tail downwards* (lacks circular tail movements of Upcher's Warbler). **Voice** Song like that of Caspian Reed Warbler in quality and rhythm, repetitive and unmusical but jaunty. Calls short *tch* or *tchek* or more drawn-out *che-ch-ch* or agitated *trrrrr*. **Habitat** Scrub, gardens, woodland, mangroves. **Note** Passage hatched.

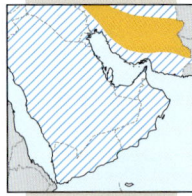

Upcher's Warbler *Hippolais languida*   PM
L: 14. Similar to Eastern Olivaceous Warbler but larger with stronger bill, and proportionately *longer, much darker sooty-brown tail with broader white tips and edges, especially visible from below. Tail uniquely waved up and down and sidewards in circular movements*. Upperparts greyer in fresh plumage (more like Eastern Olivaceous in abraded and juvenile plumage) with darker wings and prominent *pale wing-panel on closed wings in fresh plumage*. Head more rounded than Eastern Olivaceous, though pale supercilium weaker and often extends behind eye (depending on lighting); legs cold grey. Larger Olive-tree Warbler has longer wing projection, powerful yellow-orange bill and short supercilium usually only visible in front of eye. Upcher's Warbler usually prefers mid- and lower levels of bushes. **Voice** Song like Eastern Olivaceous but louder and more melodious. Call similar to Eastern Olivaceous. **Habitat** Scrub, wooded areas. **Note** Passage hatched.

Olive-tree Warbler *Hippolais olivetorum*   V
L: 16. Large, brownish-grey *with prominent wing-panel and long, dagger-like, yellow-orange bill*; forehead flat, sometimes raises crown feathers, recalling Icterine Warbler. Head pattern weak, faint supercilium *usually only in front of eye*. Underparts whitish-buff; off-white throat contrasting with rather dark head and ear-coverts. Undertail-coverts with darker 'V's, hard to see. Tail dark, edged and tipped whitish on outermost feathers; legs bluish-grey. Resembles Upcher's Warbler (q.v.). Often dips tail, sometimes waved and spread. **Voice** Song loud, *harsh and raucous*, rhythm similar to Eastern Olivaceous but slower and lower-pitched. Call is often repeated *tuc-tuc*. **Habitat** Scrub, trees. **Note** Passage hatched; vagrant Kuwait, Oman, Saudi Arabia, Yemen.

PLATE 89: ICTERINE WARBLER AND *SYLVIA* WARBLERS I

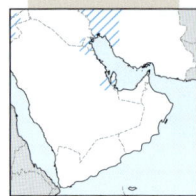

Icterine Warbler *Hippolais icterina*   V

L: 13.5. Fairly large yellow and olive warbler, showing peaked crown, heavy yellow-orange bill (red gape when singing), *very long wings with pale wing-panel in fresh plumage (less so in autumn adult) and bluish-grey legs*. Bare-faced appearance with conspicuous yellowish eye-ring and pale lores making eye rather prominent; yellowish supercilium short, rarely to rear of ear-coverts. Differs from *Phylloscopus* warblers in larger size, peaked crown, stronger bill, prominent wing-panel, and lack of contrasting supercilium and eye-stripe. Juvenile has browner upperparts and paler, more yellowish-white underparts with wing-panel more conspicuous than worn adult. Occasional individuals occur with greyish-olive upperparts and whitish underparts, recalling Eastern Olivaceous or Upcher's Warblers, but note longer wing projection and voice. **Voice** Varied fast song loud, highly mimetic, with harsh chatter and creaky 'violin' sounds, *de-de-dwiie* or *djehk-hyyii*; also includes characteristic trisyllabic *de-te-roy*; alarm call a hoarse, sparrow-like *tettettettett*. **Habitat** Trees, thickets; gardens, parks. **Note** Passage hatched, but rare; vagrant Oman, Qatar, UAE, Yemen.

Eurasian Blackcap *Sylvia atricapilla*   pm

L: 14. Similar in size to Common Whitethroat; plain grey-brown above with distinctive solidly *black cap* in adult male and *warm brown cap* in female and juveniles; lacks any white in outer tail feathers, these features preventing confusion with any other warbler. Sings from cover. **Voice** Joyful, liquid song similar to that of Garden Warbler, rich and varied, ending with a few clear melodious fluted notes. Call a hard insistent *tack*, or repetition, similar to Lesser Whitethroat; also short *churr churr*. **Habitat** Woodland, scrub. **Note** Passage hatched, occasional in winter.

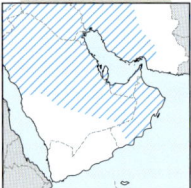

Garden Warbler *Sylvia borin*   pm

L: 14. Uniform grey-brown warbler with square-ended tail, *short, relatively heavy bill*, practically no supercilium, often greyish sides of neck, *rounded head and no white in tail or wings*. Most *Acrocephalus*, *Hippolais* and *Iduna* warblers have thinner, longer bills, *Acrocephalus* also have longer undertail-coverts and rounded tails. See juvenile Barred Warbler for separation. Usually keeps well hidden. **Voice** Song melodious, fast, with considerable jumps in pitch in irregular sequence, lacking Blackcap's ascending fluty finish. **Habitat** Woodland, scrub. **Note** Passage hatched, but scarce in Arabia.

Barred Warbler *Sylvia nisoria*   pm

L: 15.5. Large warbler, *adult with crescentic barring on underparts* (reduced in female) *and yellowish iris*; fairly long tail *has conspicuous white corners* when landing; tips of greater coverts and tertials whitish. First-autumn birds dark-eyed, recalling Garden Warbler, *but larger, with white in tail, pale edges to wing-coverts and tertials, and some dark barring on undertail-coverts*; this plumage often seen in birds the following spring. Young Eastern Orphean Warbler lacks pale markings on wing-coverts and tertials, and has contrasting dark ear-coverts, but may show dark marks on undertail. Usually furtive, keeping hidden; movements rather heavy. **Voice** Song resembles that of Garden Warbler but phrases shorter; only sub-song usually heard in Arabia. **Habitat** Thickets, scrub, trees. **Note** Passage hatched; very rare Arabia in winter.

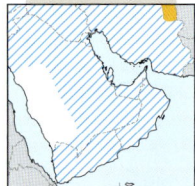

Asian Desert Warbler *Sylvia nana*   PM, WV

L: 11.5. Small, relatively long-tailed, sandy grey-brown warbler *with rufous-brown tertials, rump and closed uppertail; the spread tail is tricoloured (dark brown with white at sides and centre rufous)*, often flicked half-cocked; *iris and legs yellowish*, fine bill largely yellowish; sometimes has a pale area around eye. Terrestrial; hops and scuttles or flies low into cover, where remains hidden. Often accompanies Desert Wheatear in winter. **Voice** Call a dry, weak purring *drrrrr*, descending and fading out; also *chrr-rrr* and high-pitched *che-che-che-che*. Song, sometimes uttered in flight, starts with purring call, followed by short, clear, melodious trill. **Habitat** Desert, semi-desert, hillsides with low scattered bushes, sparsely scrubby saltflats and plains, rarely wooded areas. **Note** Passage and winter hatched.

PLATE 90: *SYLVIA* WARBLERS II

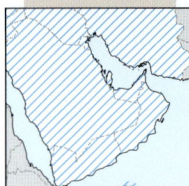

Lesser Whitethroat *Sylvia curruca* — PM, WV

L: 13.5. A small *Sylvia*, grey-brown above with browner wings, medium-grey crown and darker ear-coverts, which often vary in prominence; underparts show contrast between white throat and dusky-washed breast. Differs from Common Whitethroat in having *dark ear-coverts and dark legs* and in the *absence of rusty fringes* to wing feathers (though see *icterops* subspecies of Common Whitethroat). First-winter Lesser Whitethroat has slightly paler upperparts and often shows an indistinct whitish supercilium. Central Asian subspecies, *halimodendri*, occurs only as a migrant. It is the same size as, but paler and sandier than both *curruca* and *blythi*, having pale sandy or buff-brown upperparts, pale grey crown with clearly contrasting dark lores and ear-coverts, and more white in outer tail than *curruca*. Siberian subspecies, *blythi*, winters in Iran and parts of Arabia, and has paler, warmer brown upperparts and clean, whiter underparts, seemingly with little contrast between throat and breast. Does not songflight. **Voice** Song is a fast, loud rattle *tell-tell-tell-tell-tell*. Call (*curruca*) is a rather short and hard *tek*, often repeated; call of *halimodendri* quite different, a characteristic buzzy tit-like *che-che-che-che-che*; fast *sree-sree-sree* when agitated. **Habitat** Dense undergrowth, trees, especially *Acacia* and *Prosopis*. **Note** Passage hatched. *S. c. halimodendri* was formerly treated as a race of Desert Whitethroat *S. minula*.

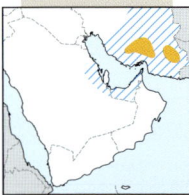

Hume's Whitethroat *Sylvia althaea* — V

L: 14. *Slightly larger than Lesser Whitethroat*, showing rather large head with *stouter bill and darker plumage*. Dark ashy-grey crown (darker than in Lesser Whitethroat), merges into dull grey-brown back; ear-coverts also dark ashy, *not contrasting with crown*, but contrasting strongly with white throat. First-winter birds can look particularly slaty above with paler feather edgings on coverts and tertials. **Voice** Song is a pleasant Blackcap-like warble. Call is a hard *tek, tek*, or a single *churrr*. **Habitat** Upland woody scrub in summer; trees and bushes on passage and in winter. **Note** Rare in hatched area in winter; vagrant Oman, Qatar.

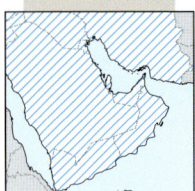

Common Whitethroat *Sylvia communis* — PM

L: 14. Medium-sized warbler, similar to Lesser Whitethroat but differs in having *chestnut fringes to coverts and secondaries*, slightly longer tail, *orangey-coloured legs* and *white eye-ring*. Females and first-winter birds lack the grey wash to head of male and have dark (not orange) iris. Subspecies occurring in region, *icterops*, has brownish (barely chestnut) fringes to the coverts and secondaries, thus resembling Lesser Whitethroat more, but can always be told by larger size and orange legs. **Voice** Song, fairly short, scratchy warbling outburst from perch on bushtop or in display flight. Call a harsh *whet-whet-whet*; loud scolding alarm is repeated drawn-out *jaairh*. **Habitat** Trees and scrub; patchy dense low vegetation. **Note** Passage hatched; rare in S Arabia in winter.

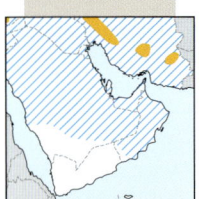

Eastern Orphean Warbler *Sylvia crassirostris* — PM, WV, mb

L: 15. *Distinctly larger than Sardinian Warbler but with grey crown contrasting poorly with blackish ear-coverts, and larger bill; adult has yellowish iris* (orange in adult Sardinian, which also has reddish eye-ring), but many singing males have dark, mud-coloured iris. Sides of tail white. First-autumn birds and immature females also *lack pale iris*; the former, as with some first-summer males, recall large, sluggish Lesser Whitethroat with contrasting dark lores and ear-coverts and dark markings on undertail-coverts. Often remains concealed. **Voice** Loud, varied, musical song recalls Common Nightingale, given from dense cover. Calls include a loud Lesser Whitethroat-like *tak* and a *trrr*. **Habitat** Bushy hillsides, deciduous thickets, parkland. **Note** Passage hatched; winters E Arabia and S Iran.

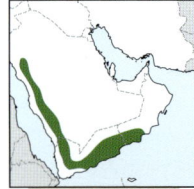

Arabian Warbler *Sylvia leucomelaena* — RB

L: 14.5. Resembles Eastern Orphean Warbler but with *shorter wings, longer tail, which is black with just thin white outer fringes* (and white tips to two outermost feathers), *unmarked whitish undertail-coverts and iris always dark*. Adult often has whitish eye-ring (lacking in Eastern Orphean); tertials fringed whitish in fresh plumage; *often flops tail downwards or in circular movements* (unlike Eastern Orphean). **Voice** Song a loud, variable slow warbling, reminiscent of Eurasian Blackcap, delivered from exposed perch; some phrases broken by drawn-out babbler-like *pift*. Calls include *tscha-tscha* and short, churring rattle *chr-rr-rr-rr-rr*. **Habitat** Acacias in semi-deserts and wadis. In SW Arabia, in rocky hills with trees and bushes; up to 1500m. **Note** Resident in S Oman. [Alt. Red Sea Warbler]

PLATE 91: *SYLVIA* WARBLERS III AND WHITE-EYES

Sardinian Warbler *Sylvia melanocephala*   V
L: 13. Slightly larger than Ménétriés's Warbler. Male distinctive with *black head, red eye-ring and contrasting white throat*. Middle East race, *momus*, shows fairly pale fringes to tertials (and is paler overall in all plumages than nominate race); otherwise with *grey upperparts and flanks*, reddish legs and much white in blackish tail. Female and first-winter have rather dark upperparts, distinctly darker than Ménétriés's Warbler. Rather skulking. **Voice** Song composed of hard call-like rattling notes mixed with short whistles in fast tempo, often given in songflight. Calls include a *sudden mechanical rattle* of 4–6 syllables *chret-tret-tret-tret-tret* or slower *terit-terit-terit-terit*; also single loud, hard *tche*. **Habitat** Dry, fairly open bushy scrub, thickets, pine and evergreen oak forests. In winter in scrub, acacia, wadis, desert-edge, oases. **Note** Vagrant Oman.

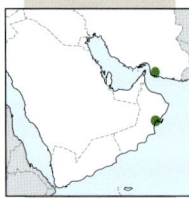

Ménétriés's Warbler *Sylvia mystacea*   PM, wv
L: 12.5. Small, *characteristically waves tail sideways or up and down*. Male has *black forehead and ear-coverts* merging into grey crown; wings with sandy fringes and black alula. In fresh plumage shows white moustachial stripe contrasting with salmon-pink to brick-red throat, paler on breast (subspecies *mystacea*; breeding E Turkey, N Iran); pink much reduced or whitish in *rubescens* (SE Turkey, W Iran, Iraq), with *turcmenica* intermediate (E Iran), but any may be whitish below when abraded. Tail dark with much white in outer feathers; *eye-ring varies from salmon-pink to red, bill bicoloured*. Legs usually pinkish-straw. *Female and young paler, of plain appearance*; with pale sandy-brown upperparts, *uniform buffish-white underparts* and base of lower mandible straw or pinkish; eye-ring from brown to yellow in adult female, dull brown to yellow-brown·in first-autumn/winter. Mostly active in low scrub or on the ground. **Voice** Song rather quiet but fast, a mixture of melodious and soft grating notes. Calls include *chak* or *tret* or distinctive *hard, quick-fire staccato rattle*; sometimes also a softer *tshshshshsh*. **Habitat** Scrub and thickets, riparian cover, often in broken country. **Note** Passage hatched; winters S Iran and Arabia.

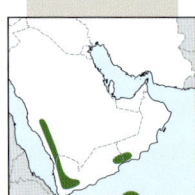

Oriental White-eye *Zosterops palpebrosus*   rb
L: 11. Small, slim and warbler-like with olive-green upperparts, prominent white eye-ring ('spectacles') surrounded by narrow dark ring, and black lores. Forehead, ear-coverts, throat and vent bright yellow, underparts otherwise off-white or greyish. Legs and rather small bill black. Forages in groups; flies erratically through or just above canopy, calling frequently. **Voice** Call weak and unremarkable but distinctive, recalling plaintive note of baby chickens or sparrows, or perhaps a distant Siskin. **Habitat** Mangroves. **Note** Recently discovered resident in Iran and Oman as mapped.

Abyssinian White-eye *Zosterops abyssinicus*   RB
L: 12. Small and warbler-like with conspicuous white eye-ring ('spectacles') and short bill. Upperparts greyish-green, throat and undertail-coverts pale yellow, breast and white belly often with smoky wash. Very active, often in pairs or occasionally small groups and constantly on the move, calling frequently. **Voice** Soft, fine and high-pitched *tiiu*; a fine purring trill or ripple; also a short low *waouw*. **Habitat** Trees in wooded mountains through to gardens from c. 300 to 3000m. Builds cup-shaped nest in outer branches of tree. **Note** Resident in S Oman; recent spread into central Oman. [Alt: White-breasted White-eye]

PLATE 92: STARLINGS I

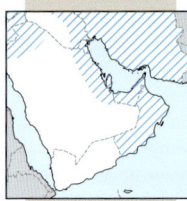

Rosy Starling *Pastor roseus* — PM, WV
L: 21. *Black-and-pink male unmistakable; long crest raised when singing*; female similar but more sombre. Juvenile lacks crest; from young Common Starling by *shorter yellow bill with culmen curved* (longer, darker, straight in Common Starling), *pale lores, paler grey-brown upperparts with whitish-grey rump and dirty whitish underparts* (Common Starling is drab brown, with dark loral streak). Gregarious. **Voice** Calls resemble Common Starling; rapid, high-pitched, musical chatter from feeding flocks. Song a lively chattering jumble mixed with melodious warbling. **Habitat** Open country, near agriculture, livestock, farmsteads. **Note** Passage hatched; some winter in Arabia. [Alt: Rose-coloured Starling]

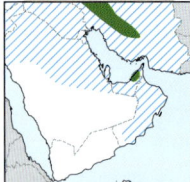

Common Starling *Sturnus vulgaris* — wv
L: 21. *Short tail and short, pointed, triangular-shaped wings with fast, straight flight* with frequent glides or brief wing-closures characteristic. *Breeding adult has glossy-green, blackish plumage with numerous minute white spots*, though spots almost absent in summer in some subspecies breeding in region, notably E Iran and Arabia, in which mantle and breast purple, and E Turkey to W Iran, which has bluish-green on mantle and bronzy sheen on head. *Juvenile drab brown, moulting to blackish with prominent white spots in first-winter*, resembling winter adult. Gregarious, roosting flocks sometimes huge; feeds aerially or on the ground; walks quickly. **Voice** Fast song, varied, whistles, strained whines and descending *seeeoo*, incorporating fine imitations. Flight call a short buzzing *tcheer*, alarm a hard *kjet*; at nest, a grating *stahh*. **Habitat** Towns, villages, farmland, woods, parks, lawns. Post-breeding, often roosts in reedbeds. Nests in hole in tree, building or nest-box. **Note** Passage and winter hatched. [Alt: Eurasian Starling]

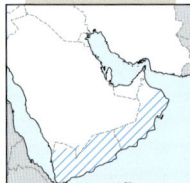

Wattled Starling *Creatophora cinerea* — pm, wv
L: 22. Resembles Rose-coloured Starling but bill stronger and wing-tip slightly blunter. Juvenile, most often seen in region, is cold grey-brown above, paler below, with pale fringes to coverts and tertials and *whitish rump*; from young Rose-coloured Starling by pale upperwing-coverts, naked malar region and lack of pale fringes to flight feathers. *Adult male has bare yellow head with black wattles*; outside breeding season resembles female, *both sexes then showing creamy-white rump, fleshy-yellow bill, buffish greater coverts, white spot on leading primary coverts, yellowish area around eye*, diffuse loral spot and moustachial streak. **Voice** Soft squeaky whistle. **Habitat** Open bush and savanna. **Note** Rare and irregular visitor to hatched area; vagrant UAE.

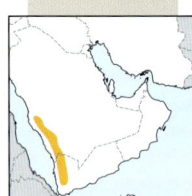

Violet-backed Starling *Cinnyricinclus leucogaster* — V
L: 19. *Male unmistakable with iridescent violet-purple upperparts, head and breast* (but can look black or red), *rest of underparts white*, eye yellow. Dark brown female has dark streaks on white breast below brownish throat, belly whiter; *eye yellow or chestnut*, but dark in otherwise similar juvenile. May flick wings singly when perched. Mostly in small groups, fairly shy. Flight direct. **Voice** Song a loud, metallic gurgling warble; call a ringing, grating musical squeal with rising inflection ending in quiet chuckle, latter also heard when flushed. **Habitat** Plains, hills and wadis with fruiting trees, mainly between 500 and 2000m. Nests in hole in tree. **Note** Vagrant Israel, Oman, UAE. [Alt: Amethyst Starling]

Superb Starling *Lamprotornis superbus* — E/I
L: 19. Unmistakable. Adults have *glossy blue upperparts and breast, green wings, whitish eyes, a narrow cream breast-band, rusty belly* and white vent and underwing. Duller-plumaged juvenile lacks cream breast-band and has dark eyes. Often swaggering and confiding. **Voice** A rising and falling chattering jumble of notes. **Habitat** Parks and gardens, open scrub, golf courses, near villages or other settlements. **Note** Not native; breeding populations originate from escapes (from Africa). In Oman, a small population may be resident near Muscat International Airport.

PLATE 93: STARLINGS II

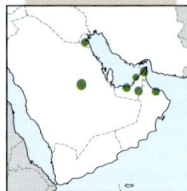

Bank Myna *Acridotheres ginginianus*   rb (E/I)
L: 21. Resembles Common Myna but smaller, with *body slate-grey* (deep vinous-brown in Common Myna); the blackish head has short crest on forecrown; *bare orange-red patch around eye and pale reddish bill diagnostic* (Common Myna has both bill and eye-patch bright yellow). *In flight, large rusty-buff patches across bases of primaries and on tail-corners*. Juveniles have body tinged brown, not slate; wing- and tail-patches more buffish-white, approaching Common Myna, but colour of bill and around eye always separates them. Often confiding, rather noisy, in pairs or in flocks; feeds mainly on the ground. **Voice** Garbled chattering notes from flock, sometimes a rather musical song, both distinguishable from Common Myna with experience. **Habitat** Towns, villages, fields, grassy areas; flocks roost in trees or reeds. Nests colonially in hole in bank, mud-well or masonry. **Note** Not native; breeding populations originate from escapes.

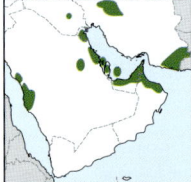

Common Myna *Acridotheres tristis*   RB (E/I)
L: 23. Bold, noisy, gregarious myna. *Body deep vinous-brown* (slate in adult Bank Myna); *bill and small bare patch below eye bright yellow* (reddish in Bank Myna). *In flight, has conspicuous large white patch across primary bases, on outer underwing-coverts and on tail-corners* (rusty-buff in Bank Myna). **Voice** Song imitative and repetitive with strident, rough and liquid notes mixed, *piu-piu-piu, che-che-che, tliy-tliy-tliy, tuu-tuu-tuu, tititi, pryv-pryv*. Alarm a grating *traaah*. **Habitat** Urban settings, parks, gardens, fields; nests in hole in tree, palm crown, under eaves or other manmade setting. **Note** Resident; breeding populations outside E Iran originate from escapes.

Chestnut-tailed Starling *Sturnia malabarica*   V
L: 21. A typical starling in shape. Adult has *whitish head*, grey upperparts and black flight feathers. *Flanks rufous-brown becoming deep chestnut on rump and outer tail feathers*. Eyes white, bill bluish at base, yellow at tip. Juvenile has whitish underparts and *chestnut tip to the tail*. **Voice** Short, harsh calls, somewhat similar to Common Starling. **Habitat** Open woodlands, farmlands. **Note** Vagrant Oman.

Brahminy Starling *Sturnia pagodarum*   pm, wv
L: 21. Small and undemonstrative, unlike Common Myna, may recall waxwing; seldom vocal, solitary or in pairs. Adult with *black cres*t, usually flattened, *rufous-buff underparts finely white-streaked and blue skin patch behind eye; bill yellow with blue base*. Vent, tail sides and tip white. Juvenile similar but much browner and lacking crest. **Voice** Often silent; short song (includes mimicry) is a drawn out gurgle followed by bubbling yodel; alarm is a Jay-like *churr*. **Habitat** Parks, gardens, woodland edge, scrub. **Note** Breeding populations originate from escapes.

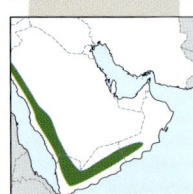

Tristram's Starling *Onychognathus tristramii*   RB
L: 25. All *glossy-black (male) or sooty-brown with grey head (female), with rusty-orange primaries conspicuous in flight*, showing as panel when perched. Usually in rather noisy flocks, but as pairs in breeding season. Often perches on camels. **Voice** Loud, echoing *'wolf-whistled' notes* heard from parties, *dee-oo-ee-o* or *o-eeou*; also mewing *vu-ee-oo*. **Habitat** Rocky hills and ravines; semi-desert, cultivations, towns and villages. Sea level to 3000m. Nests among rocks. **Note** Resident in S Oman; vagrant elsewhere in Oman. [Alt: Tristram's Grackle]

PLATE 94: THRUSHES I

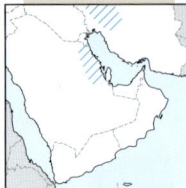

Ring Ouzel *Turdus torquatus* wv
L: 24. Resembles extralimital Eurasian Blackbird; males told instantly by *white breast-crescent and pale fringes to wing feathers*, which give 'frosty-winged' appearance in flight. Subspecies *amicorum* recorded (breeds E Turkey, Iran, Caucasus); striking, with much more extensive white in wings. Female dark brown with *obscure buffish breast-crescent* (virtually absent in first-winter females), *pale fringes to wing feathers and to feathers of underparts, giving scaly appearance*. Often shy but will also sit prominently in open. **Voice** Song highly variable, normally 3–4 fluty notes, often followed by twittering. Most common call is a loud, hard scolding *tek-tek-tek*, with a soft shrill chatter in flight. **Habitat** Mountains with alpine meadows, at upper limit of open forest. In winter/passage any area with trees or bushes; rocky hills, wadis. **Note** Partial migrant; winter hatched; vagrant Gulf States, except Saudi Arabia where rare as hatched.

Eyebrowed Thrush *Turdus obscurus* V
L: 19. Small thrush, *lacking spots or streaks on breast and belly. Male has white supercilium, black loral streak, grey head, neck, throat and upper breast, pale rufous lower breast and flanks*; female and first-autumn birds have olive-brown head, supercilium and loral streak of male, some dark streaks on whitish throat, ochre-yellowish lower breast and flanks; underwing-coverts pale grey-brown; belly white. Small whitish spots on tips of outer tail feathers in adult hard to see; first-autumn birds have white-tipped greater coverts. Flight powerful and straight, often darting low over canopy. Generally shy. **Voice** Call a thin, drawn-out *zeeip*; also a hard *dackke-dsjak, psiiie*. **Habitat** Wooded areas or open country with trees and bushes. **Note** Vagrant Oman, UAE.

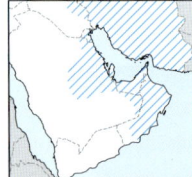

Black-throated Thrush *Turdus atrogularis* wv
L: 23. Slightly smaller than Ring Ouzel. Adult male with *clear-cut black bib, sharply defined from whitish underparts*; in female white throat streaked blackish but *lower border of black breast clear-cut*, upperparts browner. First-autumn male recalls adult but bib slightly pale-mottled; first-autumn female has *densely streaked upper breast with diffusely streaked flanks* and sometimes parts of belly; short whitish eyebrow (unlike immature Dusky Thrush). Underwing-coverts rusty, tail grey-brown above. Fairly shy. **Voice** Various calls include a quiet *chork-chork*, a quiet *sip* and squeaky *tscheeik*. **Habitat** Winters in open country with trees, grassy areas, parkland. **Note** Winter hatched, but irregular or rare; vagrant Qatar, Yemen. [Alt: Dark-throated Thrush]

Red-throated Thrush *Turdus ruficollis* V
L: 23. Adult male as Black-throated Thrush but with *red-brown bib*. First-winter male shows bright orange-brown bib. Female and immature separable from Black-throated by *rusty-rufous tail*, at least on outer feathers. First-winter female may also show slight rusty-buff on sides of streaked upper breast. Hybridisation with Black-throated Thrush can occur. **Voice** As Black-throated Thrush. **Habitat** Trees or scrub; cultivation, gardens. **Note** Vagrant Iran, Oman, Qatar.

Dusky Thrush *Turdus eunomus* V
L: 24. Size of extralimital Eurasian Blackbird. Upperparts dark olive-brown, the *closed wings having a rufous panel* (except first-autumn female); *bold white supercilium, black lores and patch on ear-coverts, white throat and lower ear-coverts, and 1–2 narrow, boldly black-flecked breast-bands, spots or flecks extending boldly along flanks*; in female and first-autumn birds white throat sometimes streaked darker, breast-bands less bold and upperparts less dark; tail dark brown; underwing-coverts pale rufous-brown. Base of bill bright yellow. **Voice** Flight calls include a drawn-out *srrii-i* and a chattering *kwae-waeg* or *tjshah-tjshah*. **Habitat** Open woodland or parkland. **Note** Vagrant Kuwait, Oman, UAE.

PLATE 95: THRUSHES II AND ROCK THRUSHES

White's Thrush *Zoothera aurea*   V
L: 28. Size and flight much as in Mistle Thrush but shyer and skulking, never far from dense cover. *Bold crescentic markings on upperparts and underparts diagnostic*; ground-colour above yellowish, below white. In flight, *bold black-and-white bands across underwing also diagnostic*, but often hard to see when flying low to cover; rump and uppertail-coverts with crescentic pattern to tail-base. Tail tricoloured with pale, whitish, corners. Usually on ground; when flushed will fly up onto branch, where may 'freeze' motionless. **Voice** Call a drawn-out *ziie*, seldom heard. **Habitat** Dense undergrowth, woodland. **Note** Vagrant Oman.

Song Thrush *Turdus philomelos*   WV, PM
L: 22. Small thrush with entirely warm brown upperparts and *blackish spots on whitish underparts* which are often washed buff on breast. In fast, straight flight shows buff underwing-coverts. In winter seen singly or in small scattered groups, often shy. **Voice** Powerful song (unlikely to be heard in Arabia) alternates between fluty and shrill sharp notes, usually repeated 2–4 times, *di-du-weet, di-du-weet, di-du-weet; dwi-dwi-dwi; du-drid-du-drid; peeoo-peeoo-peeoo-peeoo*. Call a short sharp *zit* or *zip*. **Habitat** In winter and on passage, in open country with scattered trees and bushes. **Note** Winter hatched, but rare in S Arabia.

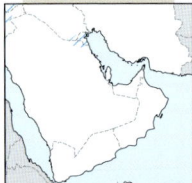

Mistle Thrush *Turdus viscivorus*   V
L: 28. Large thrush, resembling upright, outsized Song Thrush. White underwing-coverts conspicuous in flight. Upperparts grey-brown with white corners of uppertail visible on landing. Powerful flight slightly undulating and quite different from that of smaller Song Thrush. Appears pot-bellied, with stance more upright than other thrushes. Bold but often fairly shy. **Voice** Flight call a loud, dry churring *trrrrrr*, delivered repeatedly and more vehemently when alarmed. **Habitat** In winter in open country, fields with trees, grassland. **Note** Winter hatched, but rare Kuwait; vagrant Bahrain, Oman, Saudi Arabia, UAE.

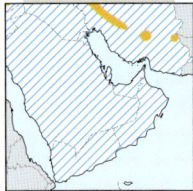

Common Rock Thrush *Monticola saxatilis*   PM, wv
L: 19. Rather small and short-tailed with longish pointed bill. Adult male easily told; in winter entire plumage edged dark and white producing scalloped effect. Female and first-winter birds browner, similar to winter male, resembling female Blue Rock Thrush but with *pale spotting on upperparts, rusty tail, and scaling on warmer buffier underparts*. Often shy and elusive. **Voice** Song similar to Blue Rock Thrush but less melancholic. Calls include a loud *chak*. **Habitat** Rocky, barren uplands; almost any habitat on passage. **Note** Passage hatched; very rare in winter Iran, S Arabia. [Alt: Rufous-tailed Rock Thrush]

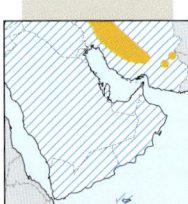

Blue Rock Thrush *Monticola solitarius*   PM, WV
L: 21. Slightly larger and longer-tailed than Common Rock Thrush and with longer bill. Male has *all-dull inky-blue plumage* (looks black at distance); in winter can show fine buffish fringes. Female dark brown, resembling female Eurasian Blackbird but has longer bill, shorter tail and dull buff spotting and barring on underparts; some show bluish tinge to upperparts. Fairly shy but will sit in full view, remaining quite still. Sings in winter quarters. **Voice** Blackbird-like far-carrying song, melancholic, the short, fluty phrases interspersed with long pauses. Calls include a hard *chak* and high *tsee*. **Habitat** Rocky deserts, mountains and cliffs; often on buildings, even in cities on passage/winter. **Note** Passage and winter hatched.

PLATE 96: ROBINS

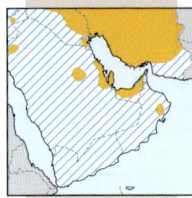

Rufous-tailed Scrub Robin *Cercotrichas galactotes*   PM
L: 15. Skulking; *long tail, often cocked and spread*. Upperparts grey-brown in subspecies *syriaca* (Turkey, Near East) and *familiaris* (Iraq, Iran, Arabia) contrasting with *rufous rump and tail*, the latter *showing prominent black subterminal-band and white tips*, obvious above and below; head distinctive with white supercilium contrasting with blackish eye-stripe. Juvenile has faintly mottled breast and flanks. Not shy but usually close to cover; often feeds on ground. **Voice** Song delivered from concealed or exposed perch or in butterfly-like songflight, is slow, clear, thrush-like, melancholic; often varied and musical, recalling lark or nightingale. Calls include a hard *teck*, a low rolling *schrrr*, and sibilant drawn-out *iiiip*; a distinctive penetrating *ssweep* when agitated. **Habitat** Semi-desert, cultivations, scrub, gardens, palm groves. **Note** Passage hatched; very rare in winter. [Alt: Rufous Scrub Robin]

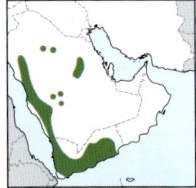

Black Scrub Robin *Cercotrichas podobe*   V
L: 18. *Entirely sooty-black*, with *prominent white tips to undertail-coverts and outer tail feathers*, readily visible when *lengthy tail is swept upwards over back* and fanned. Skulking or close to cover, often on the ground, but sings from exposed perch. **Voice** Song melodious with thrush-like whistles similar to Rufous-tailed Scrub Robin. Call a hoarse squeak, or liquid chatter. **Habitat** Desert fringe, dry scrub, cultivation, wind-breaks. **Note** Range expanding. Vagrant Bahrain, Kuwait, Oman, Qatar, UAE.

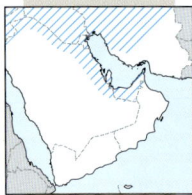

European Robin *Erithacus rubecula*   V
L: 14. Plump, upright, brown chat with *diagnostic orange-red face and breast*, large dark eyes and short wings. Tail often cocked. Usually hops on or close to ground, often shy. **Voice** Song short but crystal clear and plaintive with abrupt changes in pitch and tempo. Usual call is a sharp clicking *tic* or *tic-ik*, often repeated; also a thin drawn-out *tseer*. **Habitat** Shady gardens, copses, woodland, reedbeds. **Note** Winter hatched, but rare in south; vagrant Oman, Qatar.

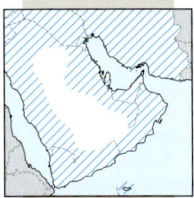

Bluethroat *Luscinia svecica*   PM, WV
L: 14. Breeding male has *blue chin and throat* framed below by black and rusty bands on breast. Subspecies assignable only in adult males: throat-patch red in *pallidogularis* and *svecica*, white in *cyanecula*; but throat entirely blue in *magna*. Non-breeding male, female and first-winter have pale throat and *black malar stripe joined to dark necklace. Rust-red sides to tail-base diagnostic*, most obvious in flight. Usually in cover on or close to ground; slinks about on foot, often darting back to cover. Tail often cocked. **Voice** Song melodic, many notes distinctly metallic or scratchy; mimetic. Calls a hard *tack*, soft *hweet*, an odd *dwzeer* and a repeated 'snipping' note. **Habitat** Mostly in reedbeds, tall grass, dense swampy cover. **Note** Passage and winter hatched; rare in north in winter.

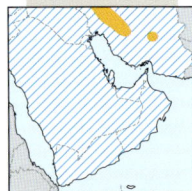

White-throated Robin *Irania gutturalis*   pm
L: 16. Size and movements recall nightingales. Striking male has *black sides of face and head framing pure white centre of throat, rusty-red underparts*, blue-grey upperparts, *black tail* and whitish supercilium. Scarcer variant with paler orange below and black line below white of chin. Grey-brown female has *dark brown tail, ochre-buff sides of body*, whitish throat bordered grey-buff at sides of head and breast. Bill long. Generally skulking, often on ground, with wings held lowered. **Voice** Song, sometimes uttered in gliding flight, fast, consists of clear whistles and scratchy harsh rolling notes mixed together; calls a nightingale-like *kerr-r-rr-rr*; wagtail-like *tzi-lit* with a hard tack in alarm. **Habitat** Stony hillsides and valleys with scrub in breeding season, usually 1,000–2,200m; on passage in scrub, woodland. **Note** Passage hatched.

PLATE 97: NIGHTINGALES AND REDSTARTS

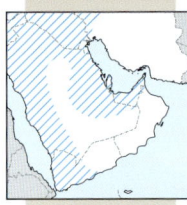

Thrush Nightingale *Luscinia luscinia* pm
L: 16. Dark olivaceous-brown, resembling small thrush, with *rusty-red tail* and pale underparts, whitish throat and eye-ring. Very similar to Common Nightingale but *darker brown above, with duller rusty-red tail and darker brownish grey breast and flanks indistinctly (but variably) mottled; far-carrying song distinctive*. Usually skulking. Unlike Common Nightingale, does not pump or wave tail. **Voice** Sings from cover, often at night. Song recalls Common Nightingale but even louder; includes *characteristic hard chucks*, dry rattles and clear whistles; lower-pitched with delivery more mechanical, less variable *djüllock… djüllock… djüllock… drlIrlIrlIrlIrlIrll-pst*, lacking distinctive crescendo of Common Nightingale. Calls include a high-pitched *hiiid* and a dry, rolling rattle. **Habitat** On passage in undergrowth, scrub, parks, gardens. **Note** Passage hatched.

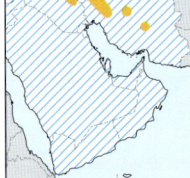

Common Nightingale *Luscinia megarhynchos* PM
L 16. Most individuals recorded in Oman referable to subspecies *golzii* (Eastern Nightingale). Very similar to Thrush Nightingale but *more russet-brown upperparts, paler, rusty-red tail*, and often more conspicuous whiter eye-ring; underparts 'cleaner', *lacking mottled impression on breast and flanks of Thrush Nightingale*. Skulking, even when singing. The subspecies *golzii* *shows pale fringes to the tertials and greater coverts, paler underparts and a pale supercilium*, with upperparts less russet, more greyish-brown. Characteristically droops wings, much like Rufous-tailed Scrub Robin; also *pumps tail downwards, and waves it sideways part-spread*, much like Upcher's Warbler. **Voice** Beautiful song, by day and night and sometimes heard on migration both spring and autumn; high-pitched comprising loud, rich, warbling whistles with distinctive crescendo, *lu-lu-lu-lu-lee-lee*, often a characteristic starting sequence, which not found in Thrush Nightingale; lacks strong chucks or frequent rattles typical of the latter. Calls similar to Thrush Nightingale include *wwheep* and a soft frog-like croak, usually from dense cover. **Habitat** Deciduous woodland, scrub, gardens, wet and dry thickets. **Note** Passage hatched; occasional in winter in S Arabia.

Eversmann's Redstart *Phoenicurus erythronotus* WV
L: 16. Slightly larger than Common Redstart. Adult male has *broad white patch on wing-coverts (including primary coverts), rusty-red mantle and most of underparts, including throat*. In winter both adult and first-winter male (which also has white in wing) have browner-grey crown and red parts fringed whitish. Grey-brown female told from similar Common and Black Redstarts by *whitish wing-bars and edges to tertials. Does not shiver tail but jerks it up and down*. Often holds wings below level of slightly raised tail. **Voice** Alarm note a croaking *gre-er*; call a loud, whistling *few-eet* and a soft *trr*. **Habitat** Oases, scrub, woodland and gardens; in summer juniper woodland and scrub. **Note** Passage and winter hatched, but rare in Arabia; vagrant Iraq, Kuwait, Saudi Arabia.

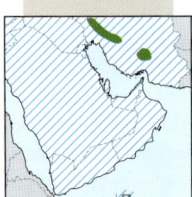

Black Redstart *Phoenicurus ochruros* PM, WV
L: 15. Subspecies *phoenicuroides* occurs in Oman. Male has *black upperparts and throat to mid-breast, sharply defined from deep red below*. Told from Common Redstart by darker upperparts, *black extending below throat* and lack of pure white forecrown. Female similar to Common Redstart but *slightly darker and drabber, particularly below* (Common Redstart more buffish below, with warmer flanks and sides of breast, and usually olive tone to mantle); some female *phoenicuroides* very similar to Common Redstart (but usually suffused rustier on lower breast and belly, with grey cast to upperparts, even if head and throat palish). Often on ground; bobs body and shivers tail. **Voice** Alarm a dry *eet-tk-tk-tk*. Distinctive song short, fast and dry *jirr-te-te-te… chill-chill-chill-chill… kretsch… sree-we-we-we*, often uttered at night. **Habitat** 2500–5000m on stony slopes, rocks, cliffs. In winter to sea level in rocky areas, open woodland and villages. **Note** Passage and winter hatched.

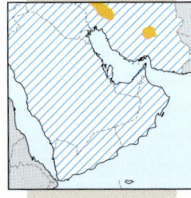

Common Redstart *Phoenicurus phoenicurus* PM
L: 14–15. Male with *black cheeks and throat contrasting sharply with rusty-red breast* and belly; crown grey with *pronounced white forehead*. Male *samamisicus* (Ehrenberg's Redstart, breeds Turkey eastwards; a few recorded on passage early spring) *has white wing-panel* and, often, darker upperparts; autumn male Common Redstart has black areas fringed pale. Female brownish-olive above, separated from similar Black Redstart by *paler, warmer buffish-white underparts*. Hunts insects in flycatcher-like fashion; bobs body and shivers tail (as Black Redstart). **Voice** Call resembles Willow Warbler's soft *wheet*, often followed by *tuuk-tuuk*. Song short and melodious, *seeh-truee-truee-truee-see-see-seeweh*; frequently imitates other birds. **Habitat** Woodland, parks and scrub on passage. **Note** Passage hatched.

PLATE 98: CHATS

Pied Bush Chat *Saxicola caprata* — V
L: 13. Slimmer and slightly longer-tailed than other stonechats; tail flicked less. *Jet black male has white belly, rump and narrow shoulder-patch* easily seen in its low jerky flight. Female from other chats by *unstreaked sooty earth-brown upperparts and breast*, creamy belly and rufous-orange rump, some being more rusty-brown on breast and having a slight supercilium. **Voice** Song a short rich warble of whistling notes; alarm a curt *chuk* and a *chek-chek-trweet*. **Habitat** Cultivation, scrub, marshes. **Note** Some dispersal; vagrant Oman, Qatar, Saudi Arabia, UAE. [Alt : Pied Stonechat]

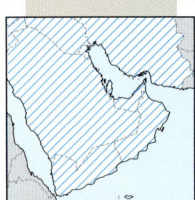

Whinchat *Saxicola rubetra* — PM
L: 12.5. Short-tailed chat with slightly smaller head and longer primary projection than European Stonechat, reaching almost halfway down tail. From stonechats by *combination of streaked brownish rump and white sides to base of tail*. Male has bold, clear-cut white supercilium and white stripe between blackish sides of head and orangey throat; female duller with paler throat and browner sides of head. *White spot visible on primary coverts*, particularly in flight (almost absent in some first-autumn birds). Confusion possible in autumn with immature Siberian Stonechat. **Voice** Lilting song short, fast and abrupt, variable, usually a mixture of melodious and scratchy notes; often imitates other birds; alarm call *djü-tek-tek*. Silent on passage. **Habitat** Open country, marshes, scrub; on passage in fields and other open ground. **Note** Passage hatched.

European Stonechat *Saxicola rubicola* — WV
L: 12. Short-tailed, short-winged chat with large rounded head and upright stance, frequently flicks wings and tail. Adult male easily told from Whinchat *by black head and throat, reddish breast, white neck-patch and black tail*. Female duller with dark brown head and usually throat, reddish-brown upper breast, with almost Whinchat-like supercilium variably present in autumn. Male European Stonechat has variable width grey-brown to white rump; if latter usually (but not always) narrow or with some dark streaks; *grey underwing-coverts and axillaries* (blackish in Siberian Stonechat) with white collar narrower and less extensive than in male *maurus* Siberian Stonechat. Female European has dark earth-brown back, usually a *streaked brownish rump*, and is dusky or whitish on throat (Siberian typically only whitish), although some perhaps indistinguishable from *maurus*. Female Siberian Stonechat has paler sandy to warm buff upperparts, often with a more obvious supercilium, and a pale unstreaked rump. Immature European Stonechat is darker, lacks pale rump of *maurus* and has less extensive white on coverts. **Voice** Short song has irregular, rapidly repeated series of double notes; alarm call *wheet-trak-trak* like pebbles hit together. **Habitat** Open terrain; sea level to over 3,000m, in cultivated areas or scrub-covered slopes. **Note** Breeding range apparently expanding eastwards; winter hatched, but rare in Arabia.

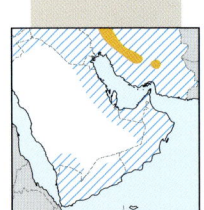

Siberian Stonechat *Saxicola maurus* — WV, PM
L: 12. Male in breeding plumage has broader, more extensive white collar than European Stonechat and *conspicuous, broad, white to salmon-pink or orange-buff rump*. First-autumn Siberian Stonechat superficially recalls autumn Whinchat. Female lacks the clear-cut, long, creamy-white supercilium of Whinchat, but often difficult to separate from European Stonechat (see above). Males of different subspecies described below.
S. m. maurus (breeds European Russia) has longer primary projection than short-winged European Stonechat (of W and central Turkey); male *maurus* has *unstreaked orange-buff to white rump* (wider than in European male); *black tail*; broader white half-collar; *paler back and axillaries jet black* (greyish-white in male European Stonechat).
S. m. variegatus (formerly *S. m. armenicus*, breeds SE Turkey, N and W Iran) is also long-winged but has less white at sides of tail (not always visible) than *hemprichii*; male has dark chestnut breast contrasting with pure white belly.
S. m. hemprichii (Caspian Stonechat; formerly *S. m. variegatus*, breeds Caspian) has even longer primary projection (approaching Whinchat); *much white in sides of tail-base* (recalling Northern Wheatear); the palest subspecies, male with warm buff upperparts and large white patch on rump, sides of neck and shoulders.
Voice As European Stonechat. **Habitat** Open terrain; sea level to 3,000m, in cultivated areas or scrub-covered slopes; on passage/winter any open ground. **Note** Winter hatched. Distribution poorly documented by subspecies; *maurus* most frequently reported in region, *hemprichii* scarce or vagrant in Arabia including Oman; winter range of *variegatus* unclear – likely to occur, but no accepted records for Oman.

PLATE 99: WHEATEARS I

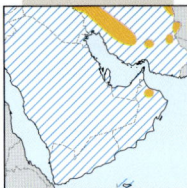

Isabelline Wheatear *Oenanthe isabellina*   WV, PM, mb

L: 16. *Resembles large female Northern Wheatear, usually more robust in build*, with longer legs and apparently also bill, slightly shorter tail and more upright stance. *Best separated by isolated black alula in paler, more uniform sandy wings, contrasting less with upperparts*, broader pale feather-edges to wings than Northern Wheatear (some first-autumn Northerns can show pronounced isolated dark alula); *broader black terminal tail-band with shorter stalk to 'T'. Supercilium usually broadest and whiter in front of eye* (in Northern narrower and buffier in front of eye). In low flight shows *half-translucent, dark-tipped primaries; underwing-coverts and axillaries buffish-white* (dusky-grey in female Desert Wheatear; dark grey, broadly tipped whitish in Northern Wheatear). Runs mostly; wags tail strongly and frequently. **Voice** Song, often in display flight, longer and more variable than Northern Wheatear with less scratchy notes and often ending with series of whistles; mimics well. Alarm call *tjack-tjack*, sometimes followed by slightly descending *hiu* or *diu*. **Habitat** Barren or grassy areas, steppes; on passage/winter also cultivation. **Note** Passage and winter hatched; absent from north in winter.

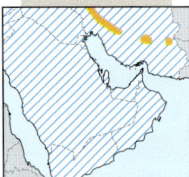

Northern Wheatear *Oenanthe oenanthe*   PM

L: 15. Relatively short-tailed wheatear with *blackish terminal tail-band of even width*. Male readily told by ash-grey crown and back, white supercilium, thin black eye-stripe and black ear-coverts. Female is greyer-brown than similar Eastern Black-eared and Pied Wheatears; first-autumn birds *best told by shorter tail* (wing-tips closer to tail-tip) *and tail-band of even width*. In Eastern Black-eared and Pied tail longer, wings half or less of tail, which has more white and narrower black tail-band with black extending upwards on outer tail feathers. See also Isabelline Wheatear. Restless; bobs body, wags tail and flicks wings. **Voice** Song short with fast chacking call-notes mixed with high-pitched whistles in irregular rhythm. Alarm a hard *tack-tack* or *hiid, tack-tack*. **Habitat** Uplands with rocky or stony slopes, often with bushes; on passage to sea level in any open area, including cultivations. **Note** Passage hatched.

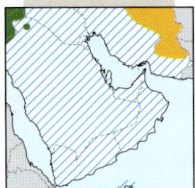

Desert Wheatear *Oenanthe deserti*   WV, PM

L: 14.5. Easily told in flight from all other wheatears in region *by almost wholly black tail* (no white at sides). Male recalls black-throated form of Eastern Black-eared Wheatear, *but black throat joins to narrower black of wings* (often with whitish scapulars contrasting); in flight shows whitish wing-panel and white rump *tinged buffy towards top*. Sandy-brown or grey-buff female often lacks black throat; may recall female Eastern Black-eared but always told by rump and tail pattern. Often confiding. **Voice** Short piping song plaintive, with downward inflection, occasionally includes rattling notes; calls include a soft whistle. **Habitat** Shrubby desert, barren stony areas. **Note** Passage and winter hatched.

Finsch's Wheatear *Oenanthe finschii*   V

L: 15. Heavier and stockier than Pied and Eastern Black-eared Wheatears; male told from former by *narrow creamy-buff (milky tea) or, when worn in spring, silvery-white stripe down mantle and back to join white rump*; from latter by *larger black 'bib' broadly connected with black wings* (beware Black-eared with head sunk between shoulders); *also from both by terminal tail-band of even width* (no black extension up sides) *and pale greyish flight feathers below, appearing translucent above*. Crown as mantle, but centre flecked dusky when worn. Some females have variable blackish on throat (sometimes lower throat only); pale-throated birds told from female Eastern Black-eared/Pied by *sandy brown-grey upperparts*, contrasting with darker, browner wings, creamy breast lacking orange-buff tone of female Eastern Black-eared, paler flight feathers below and tail pattern. *Flight feathers and primary coverts often finely pale-tipped* (in both sexes into second calendar year). Female often shyer than male. Ground-dwelling, perching infrequently in trees. Frequently *bows low, cocking tail, repeatedly spreading and lowering it slowly*. Has descending zig-zag songflight. **Voice** Song short and rich with scratchy notes often mixed with clear whistles, phrases intermittent, including musical *ctsi-tsi-tseeoo*. Alarm call *tack*; also *che-che-che*. **Habitat** Dry rocky, stony uplands and foothills, sparsely vegetated semi-deserts. **Note** Passage and winter hatched; rare in Arabia; vagrant Oman.

PLATE 100: WHEATEARS II

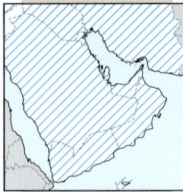

Pied Wheatear *Oenanthe pleschanka* PM
L: 14–16. Slender, often perches on bushes. *From Eastern Mourning and male Arabian Wheatears by absence of white panel in open wing*. White-throated form, '*vittata*', similar to Eastern Black-eared Wheatear but mantle black. Autumn male has black back and throat fringed buffish, dark crown with buff-white supercilium, and buff underparts. Female like female Eastern Black-eared Wheatear, but *upperparts usually duller, colder brown-grey*; some show large dark greyish 'bib' in summer (absent in female Eastern Black-eared); autumn female also told by dark brown breast-sides merging with greyish 'bib'. First-autumn female told with difficulty from Eastern Black-eared by colder tone above; crown, mantle and shoulders usually *scalloped with rows of pale fringes* (absent or ordered erratically in Eastern Black-eared). First-winter female from Northern Wheatear by *black extending up sides of tail*; the black tail-band *sometimes of uneven width*. **Voice** Short musical song, often in flight; twittering phrases resembling lark or wagtail; often mimics. Calls hard *tack*, dry *trrrlt* or dry, sneezed *snerr*. **Habitat** On passage and winter in rocky terrain, bare fields, wasteground. **Note** Passage hatched; occasional in winter in S Arabia. Form '*vittata*' regular in small numbers in Oman in spring.

Cyprus Wheatear *Oenanthe cypriaca* V
L: 13.5. Sexes closely similar; resembling male Pied Wheatear but warmer buff below black 'bib'. *Best told by song, narrower white rump, shorter primary projection* and smaller size. Female has slaty to grey-brown upperparts and rather dark crown (all white in male), surrounded by whitish stripe. Adult in autumn has grey-buff fringes to black mantle and throat, dark crown, buff-white supercilium and *underparts deeper rusty-buff than Pied Wheatear*. First-autumn birds similar but with more, and broader, pale-fringing. In winter underparts rapidly bleach paler, thus differing little from Pied Wheatear. **Voice** Song *recalls cicada* but less harsh, a lengthy purring *bizz-bizz-bizz*, often ending in high-pitched, drawn-out piping note, sometimes uttered in flight. Calls include a hard *tack*. **Habitat** As Pied Wheatear, but often on more forested slopes; on passage in wadis, scrub and cultivation. Nests in hole in bank, even a nest-box. **Note** Vagrant Oman.

Black-eared Wheatear *Oenanthe hispanica* WV, PM
L: 13.5–15. Subspecies *melanoleuca* (Eastern Black-eared Wheatear) occurs. Small, build much as Pied Wheatear and sharing same (variation in) tail pattern. Male of black-throated form can be confused with male Finsch's Wheatear but *black of throat/ear-coverts not joined with black of wings and shoulder*; mantle whitish (summer) or buffish-grey (autumn). White-throated male told from '*vittata*' form of Pied Wheatear by pale mantle. Female Eastern Black-eared difficult to separate from Pied Wheatear, but has whitish chin (dusky in Pied), usually with slight rustiness below and *sandier mantle*. **Voice** Song resembles Pied, rather variable, dry, scratchy. Calls include a hard *tack*, sneeze and a characteristic buzzing (like an angry fly!). **Habitat** Sparsely vegetated, stony slopes; any open area on passage. **Note** Passage hatched.

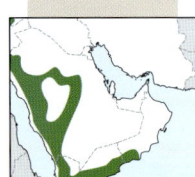

Blackstart *Oenanthe melanura* RB
L: 15. Slender, relatively long-legged chat with *all-black tail which is slowly lowered and spread, often coinciding with half-spreading of wings*. Nominate subspecies (N Arabia, Near East) is pale ash-grey above, whitish-grey below with whitish wing-panel; S Arabian *erlangeri* almost uniform smoky-grey above, underparts little paler; wing-panel brownish. Perches freely on low branches or rocks, typically flirting wings and fanning tail; often inquisitive and approaching observer closely. **Voice** Short, mellow, subdued, simple song, sometimes uttered in flight, an often repeated *che-we-we* or *ch-lulu-we*. Alarm a short, deep *tjaet-aeteh*. **Habitat** Sparsely scrubby slopes, cliffs and bare rocky wadis. **Note** Breeding resident in S Oman; vagrant UAE. [Formerly in genus *Cercomela*]

PLATE 101: WHEATEARS III

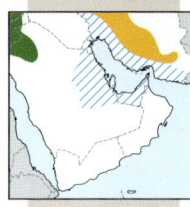

Mourning Wheatear *Oenanthe lugens* — pm, wv

L: 13.5. Iranian subspecies *persica* (Eastern Mourning Wheatear) occurs Oman, UAE; similar *lugens* occurs Near East/NW Saudi Arabia. Sexes similar; resembles stocky male Pied Wheatear when perched, *but undertail-coverts apricot (all ages); in flight shows prominent whitish wing-panel.* Compared with Pied Wheatear the black bib is smaller, underparts whiter (tinged buff in Pied), primary coverts narrowly tipped white and black band on tail lacking the black extension up the outer feathers of Pied. In autumn lacks the pronounced pale feather fringes to black throat and mantle of male Pied. **Voice** Song a lively twitter; call *check-check*; alarm *peet-peet*. **Habitat** Coastal and inland bluffs in desert and semi-desert in winter, sometimes near remote habitation. **Note** Winter hatched, but scarce in Arabia.

Arabian Wheatear *Oenanthe lugentoides* — RB

L: 13.5. Resembles Eastern Mourning Wheatear (breeding ranges do not overlap) in having apricot/rufous-buff undertail-coverts. Male Arabian has slightly more extensive black on throat, sides of neck and back, narrower white rump-patch, less extensive white crown, *often streaked grey* (sometimes crown grey with white sides); underparts below 'bib' as Eastern Mourning, but outer tail feathers with a little more black. *In flight, shows a small, but conspicuous white primary patch* (smaller than in Eastern Mourning; absent in other wheatears). Female charcoal- to grey-brown above with warm orangey-brown ear-coverts (sometimes entire head); *breast often diffusely grey-streaked*; lacks white primary patch of male, but primaries may appear silvery-grey at base. **Voice** Song short, loud musical bubbling; also musical *too-too*. Calls *chuck-a-doo* (like stones knocked together), Tree Sparrow-like *tek-tek*, rasping *kaak* often repeated and interspersed with high-pitched *seeek*. **Habitat** Rocky hillsides, mountains with sparse vegetation, juniper scrub near cultivation; usually 1000–2500m, occasionally down to sea level. Nests in hole in rocks, wall or bank. **Note** Subspecies *O. l. boscaweni* occurs. [Alt: South Arabian Wheatear]

Hume's Wheatear *Oenanthe albonigra* — RB

L: 16.5. Sexes alike. Closely resembles *picata* subspecies of Variable Wheatear, but *larger, with 'bull-headed' appearance, velvety black plumage more glossy and white on back extending farther up between wings, where border to black mantle is rounded* (square-cut in Variable Wheatear). Black throat has slight side extension thus less 'bib-shaped' than Variable Wheatear. Underwing less contrasting, but tail pattern like Variable Wheatear. Juvenile like adult, but plumage matt blackish-brown (with yellow gape line). Stance more upright than in Variable Wheatear; prefers open rocky terrain, only rarely on plains or in more wooded areas. Often inquisitive, at other times shy. On territory year-round, but sometimes descending from higher altitudes. **Voice** Loud, melodious ringing song, a short lyrical jumble, recalling Whinchat. Ventriloquial. Call sharp, short and high-pitched; alarm harsh and grating. **Habitat** Mountain slopes, foothills down to sea level and boulder-strewn barren hills with scant vegetation; often on buildings or overhead wires. Nests in hole in rock or scree. **Note** Winter dispersal hatched; vagrant Bahrain, Kuwait.

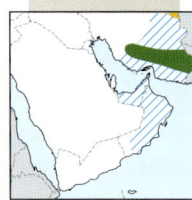

Variable Wheatear *Oenanthe picata* — PM, WV

L: 15. Three subspecies occur, but only *picata* likely to be seen in region, the other two (*capistrata* and *opistholeuca*) being rare or vagrant to Iran. Subspecies *picata* (breeds Iran, winters south to UAE and Oman): resembles small Hume's Wheatear but *more slender. Male has crown, upperparts and 'bib' dark charcoal (matt, without gloss, unlike blacker Hume's)*; rest of underparts white, undertail-coverts sometimes buff in autumn. Female usually matt sooty-brown above where male charcoal blackish, 'bib' often rufous-tinted but throat sometimes blackish; rest of underparts creamy-buff to white. *Perches low, in winter often in trees (like redstart) rather than on rocks.* **Voice** Song rather scratchy warble; mimetic. Ordinarily silent in winter. **Habitat** Barren, boulder-strewn country; hillocks with sparse woody vegetation, steep riverbanks; in winter arid stony plains with trees and outcrops, cultivation. **Note** Passage and winter hatched (subspecies *picata*), but mostly scarce in Arabia; vagrant Saudi Arabia. [Alt: Eastern Pied Wheatear]

PLATE 102: WHEATEARS IV

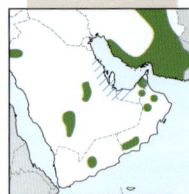

Hooded Wheatear *Oenanthe monacha*   pm, wv, rb
L: 17. *Large, slender, long-tailed and long-winged wheatear with long bill and buoyant, almost butterfly-like flight*, recalling Spotted Flycatcher when catching prey, sometimes in long sallies. Sexes differ. Adult male distinctive, with *whitish crown, black below extending to centre of breast* and, except for black central tail feathers, *nearly all-white tail with just black corners*. Autumn and juvenile males have creamy-buff crown, whitish fringes to black throat, wing-coverts and mantle, with lower underparts, rump and sides of tail tinged buffish. Female sandy brownish-grey above, *merging into cream-buff rump, tail-coverts and sides of tail*, in which central feathers and tail-corners are dark brown; whitish-grey underparts washed buff at sides of breast, flanks and undertail-coverts. In autumn female and juvenile, the rump and underparts may appear reddish-buff with almost reddish-brown sides of tail, but absence of dark terminal tail-band separates from Red-tailed Wheatears. **Voice** Song has short melodious phrases, interspersed with some stone-clicking notes; brief throaty thrush-like warble heard infrequently, relatively simple, lilting and slightly sad. Female utters a *whit-whit* or repeated *jiirp* like a fledgling Eurasian Blackbird; also a *wit-awheet-wheet-wheet* or *whee-whee-whee-wheeoo*. **Habitat** Desolate, barren rocky ravines, gorges and deserts. Nests in hole in rock. **Note** Winter dispersal hatched.

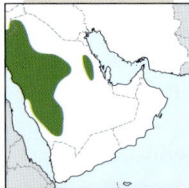

White-crowned Wheatear *Oenanthe leucopyga*   V
L: 17. Large wheatear with sexes similar; glossy *black with black underparts down to legs, and, in many adults, a white crown*; immature and some adults have black crown but size, long bill *and white sides of tail with black corners diagnostic* (no black terminal band). Some black-crowned birds show a few white feather-tips, eventually developing a white crown. In male Hooded Wheatear, which has similar tail pattern, black below extends only to centre of breast. **Voice** Variable song has whistling and tuneful notes, sometimes scratchy, often with imitation of other locally occurring species; common phrase *viet-viet-dreeit-deit*, slightly descending but much variation. Call *peeh-peeh*. **Habitat** Rocky deserts, ravines in rocky mountains, usually without vegetation; often around human settlement. **Note** Some winter dispersal, rare Kuwait; vagrant Bahrain, Oman, Qatar, UAE.

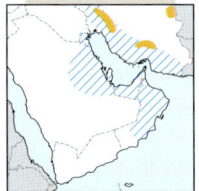

Red-tailed Wheatear *Oenanthe chrysopygia*   WV, PM
L: 15. Sexes alike; *upperparts drab greyish, especially head, face pattern rather bland (like autumn flava wagtail), rump and sides of tail orangey-rufous* (rump sometimes paler); *tip of tail fringed rufous when fresh; underwing coverts pale/off-white* (dusky-grey in Kurdish Wheatear), *vent and flanks rufous-orange* (paler orange to buff or whitish in female Kurdish Wheatear). Long slim bill; silhouette can recall Blue Rock Thrush. **Voice** Warbling song loud with adept mimicry. Undemonstrative and ordinarily silent in winter; occasionally giving grating alarm. **Habitat** Stony or barren hillsides, low scrubby vegetation. In winter cultivation, ruins, rubble and dumps in sand desert. **Note** Passage and winter hatched.

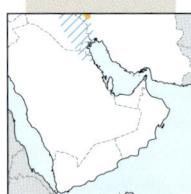

Kurdish Wheatear *Oenanthe xanthoprymna*   V
L: 15. Male distinctive, with whitish supercilium, black throat, sides of head and neck merging with blackish-brown wing-coverts, *rufous rump, with white sides to tail-base in adults (rufous in immatures)*; black band at tip of tail narrow; *little white in wings*. Female confusable with Red-tailed Wheatear (which see), though sometimes shows dark throat. Usually solitary; has bounding hops and slight downward tail-flicks; often flies with tail closed. **Voice** Brief song a slow throaty warble. Calls include *steu-steu-steu*; alarm note a short dry *zuk* or *zvee-tuk*. **Habitat** Favours rocky outcrops. In winter, hillsides, cultivation, ruins. **Note** Winter hatched, but distribution poorly known; vagrant Oman, UAE. [Alt: Kurdistan Wheatear, Rufous-tailed Wheatear]

PLATE 103: FLYCATCHERS I

Spotted Flycatcher *Muscicapa striata* PM
L: 14. Brownish-grey, long-winged flycatcher *with streaked forehead, crown and breast* (though may not be obvious), *variable pale edges to greater coverts and tertials* (depending on subspecies and wear), *and blackish bill* and legs. Faint whitish eye-ring. Sexes similar. Eastern subspecies *neumanni* is distinctly paler and greyer above than nominate *striata* or *inexpectata*. Makes short aerobatic flights (may hover) from exposed branch to catch prey, often returning to same perch. Perches upright, often flicking wings. **Voice** Call a sharp *tzeet*; alarm *isst-tek*. **Habitat** Gardens, parks, woodland. **Note** Passage hatched.

Asian Brown Flycatcher *Muscicapa dauurica* V
L: 12.5. A small, rather featureless, compact flycatcher, like a poorly marked, diminutive Spotted Flycatcher. *Upperparts unstreaked grey-brown*, underparts off-white with *grey wash across breast; white eye-ring* and white lores are fairly obvious features. Bill strong and broad with pale base to lower mandible. In first-winter, secondaries and tertials are edged pale, and greater coverts are tipped pale. Behaviour as Spotted Flycatcher. **Voice** Piercing *tzi* occasionally given by migrants. **Habitat** Open woodland, trees. **Note** Vagrant Oman.

Red-breasted Flycatcher *Ficedula parva* pm, wv
L: 12. Small flycatcher with conspicuous white patches at sides of tail-base, whitish eye-ring, and straw or yellowish base to bill (when seen well); adult male has *reddish-orange throat and sometimes upper breast* and lead-grey sides of head and neck; female and second-year male have buffish-white throat. Often cocks tail. See similar Taiga Flycatcher below. **Voice** Call a dry, rolling *terrrr*; also a thin *tsri*, a *tek* or, in alarm, *tee-lu*. **Habitat** On passage/winter anywhere with trees or scrub. **Note** Passage hatched; winters in SE Arabia.

Taiga Flycatcher *Ficedula albicilla* V
L: 12. Very similar to Red-breasted Flycatcher; differs in having *bill all dark or barely paler on base of lower mandible* (straw- or yellowish-based in Red-breasted), *coal-black uppertail-coverts and different call*. Female, as with winter male and juvenile, colder less buff below than Red-breasted, again distinguished by dark bill and rump, and call. Although not diagnostic, first-winter Taiga also typically shows broad, cold white fringes on the outer webs of the tertials, broadening and expanding to a round blob at the shaft (as first-winter European Pied Flycatcher), whereas in Red-breasted both webs have warmer yellowish-buff fringes, with a thorn shape at the shaft. **Voice** Call is *an insect-like buzz* rather than dry roll of Red-breasted. **Habitat** Wooded areas. **Note** Vagrant Oman, UAE.

European Pied Flycatcher *Ficedula hypoleuca* V
L: 13. Male black-and-white *with black hindneck,* small white forehead-spot (often divided in centre) and *narrow white streak at base of primaries* (sometimes absent). Some males (rare variant) are grey-brown above but still with white spot on forehead. First-winter has dark tertials, white-bordered on outer webs, with rounded blob at end and only *small white streak at base of primaries*. For separation from Semicollared Flycatcher, see under species. Often flicks wings. **Voice** Calls on passage *tuk*, and a short metallic *twink*, different from call of Semicollared. **Habitat** Trees and woods. **Note** Passage hatched; vagrant Oman, UAE.

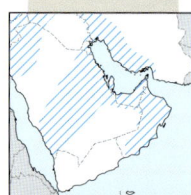

Semicollared Flycatcher *Ficedula semitorquata* V
L: 13. Resembles European Pied Flycatcher; also flicks wings. Male told by *white half-collar onto sides of neck, larger white spot at base of primaries* (absent, or just a narrow streak in European Pied), more white at sides of tail and white-tipped median coverts (rarely so in European Pied); extent of white in rest of wing and on forehead is greater than in European Pied. Female greyer above than European Pied; *often* (but not always) *more white at primary bases and on tips of median coverts (as well as greater coverts)*; however, this second wing-bar on median coverts is not reliable for first-autumn birds. **Voice** Call single dry *thuk* (close to Red-breasted Flycatcher); also hard *tack*, alarm call *eeet*. **Habitat** Woods, parks, large gardens. **Note** Passage hatched.

PLATE 104: FLYCATCHERS II AND SUNBIRDS

Blue-and-white Flycatcher *Cyanoptila cyanomelana*   V
L: 17. Larger, longer-tailed and heavier-billed than Spotted Flycatcher. *Male dark blue above, with shining blue crown, bluish-black sides of head, throat and breast, sharply demarcated from white belly; white sides to tail-base* usually visible in flight only. *Dark olive-brown female is paler below with distinct creamy throat-patch*, white belly and undertail-coverts; indistinct pale eye-ring; first-winter male has *bright blue rump and tail, latter tipped black, but white sides at base*; wings bright blue with black-tipped primaries. Upright stance with frequent slow tail movements and wing-flicks; swoops to ground to feed, then returns to perch. **Voice** Grating *tchach* or *tek-tek*. **Habitat** Any area with trees. **Note** Vagrant Oman, UAE.

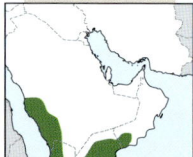

Nile Valley Sunbird *Hedydipna metallica*   rb
L: 10 (breeding male with full tail 15cm). *Small with short, slightly decurved bill, which separates it from all other sunbirds in Middle East. Male has yellow underparts with dark purple and green upper breast and throat; note elongated tail feathers in breeding season*. Female and juvenile grey-brown above, with throat whitish and *underparts washed yellow* (separating it from similar Palestine and Shining Sunbirds). Non-breeding male recalls female but underparts yellower; some have blackish centre of chin and throat. Often flicks wings and tail. Nomadic. **Voice** Song soft and high-pitched with trilling and hissing notes, *pruiit-pruuiit-pruuit-tirirriri-tiririri*; also thin hoarse *veeii-veeii*, a *ptscheeciii*, repeated *cheeit-cheeit* and *tee-weee* with upward inflection on last note. **Habitat** Gardens, savanna, wadis with dry scrub, from sea level to 2500m, commonest at low altitudes. Oval nest is suspended from twig. **Note** Resident S Oman with some post-breeding dispersal.

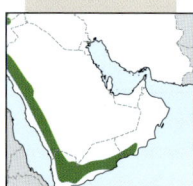

Palestine Sunbird *Cinnyris osea*   RB
L: 11. Small with medium-long decurved bill. *Male told from Shining Sunbird by bluer plumage, smaller size, slightly shorter bill, absence of red breast-band and quicker flicks of shorter tail. Female distinctly paler grey below than larger female Shining, ventral region plain, off-white.* From Purple Sunbird by different distribution, *slightly longer bill; male with bluer gloss in Palestine (greener in Purple Sunbird) and in female lack of yellow below*. Flight rapid with irregular, short, dipping undulations. **Voice** Fast song a high-pitched trilling *dy-vy-vy-vy-vy-vy* or rising *tweeit-tweeit-tweeit*, accelerating and often ending in European Serin-like trill. Calls include a hard Lesser Whitethroat-like *tek* or chittered *tek-tek-tek*, loud sweet *doo-swee* or *dee-swit*, repeated strained *cheeooo* and a sharp *te-veeit, te-veeit*, the second note stressed and rising. **Habitat** Well-vegetated areas, including rocky wadis, plains with acacias, riverbanks, gardens, from sea level to 3200m. Nest suspended on tree or bush. **Note** Resident S Oman.

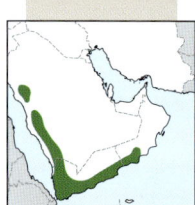

Shining Sunbird *Cinnyris habessinicus*   RB
L: 13. Medium-sized *with long decurved bill. All plumages much darker below than other sunbirds in region*. Male bright metallic-green *with broad, though inconspicuous, reddish breast-band (absent in male Palestine Sunbird). Female dark sooty-grey, slightly paler below, but much darker than female Palestine; feathers of ventral region fringed whitish*. Juvenile male resembles female but has black centre of throat and often a dark breast-patch, whitish fore-supercilium and moustache. Non-breeding male like female but with some green on head, back and breast. Has rather pronounced, slow flicking of longer, broader tail than Palestine and flight has longer, deeper undulations. **Voice** Fast song fluty and trilling *tuu-tuu-tuu-tuu-vita-vita-vita-du-du-du-du*, often ending in Winter Wren-like trill. Sub-song a fast whispering warble. Calls include a *distinctive loud winnowing whistle* and a hard *dzit*. **Habitat** Luxuriant vegetation in wadis; dry savanna scrub, 250–2500m but mostly lower and middle altitudes. Bottle-shaped nest suspended in tree or bush. **Note** Resident S Oman.

Purple Sunbird *Cinnyris asiaticus*   RB
L: 10. Resembles Palestine Sunbird *but bill slightly shorter, less decurved and has different distribution*. Metallic bluish-black *male sometimes has a narrow red-brown breast-band*; mousey-brown *female has underparts washed yellow* (pale greyish in female Palestine). Male in eclipse (Sep–Dec) like female but underparts yellower with dark line down centre of throat and breast. Short-tailed, hummingbird-like appearance, feeding largely on nectar from flowers, with corresponding seasonal movements. **Voice** Excited song repeated two to six times, *cheewit-cheewit...*, with Willow Warbler-like cadence. Male's sub-song is a low twitter; calls *dzit-dzit* and pronounced *tsweet*. **Habitat** Gardens, cultivation, tamarisks along rivers, thorn scrub, dry forest, stony desert with flowering trees and shrubs (acacias, Sodom Apple *Calotropis procera*), mangroves. Hanging nest is pear-shaped, suspended in tree or bush. **Note** Resident N Oman and UAE. Vagrant Kuwait.

PLATE 105: SPARROWS

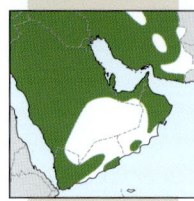

House Sparrow *Passer domesticus* RB
L: 14.5. Subspecies *indicus* (Indian House Sparrow) is the taxon that occurs in Arabia. *Male has grey crown, chestnut-brown sides of head, variably large black 'bib', very pale, clean underparts, strikingly white cheeks*, with prominently dark streaked back. Female and juvenile buffish-brown above, *greyish on crown, boldly streaked darker on mantle* and pale greyish below; supercilium creamy. Feeds and roosts in flocks. **Voice** Simple song monotonous, cheeping or chirping notes of varying pitch. **Habitat** Towns, villages, farmland. Colonial; untidy nest built in rock crevices, buildings, trees or bushes. **Note** Resident in much of Arabia east to SE Iran. Race *hufufae*, sometimes included in *indicus* 'group', occurs E Arabia. Map includes all House Sparrow races.

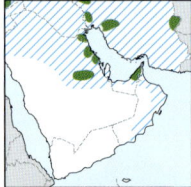

Spanish Sparrow *Passer hispaniolensis* pm, wv
L: 15. Resembles House Sparrow (with which it hybridises); larger bill sometimes evident. *Male has rufous-brown crown (pale-fringed duller in winter) and large black 'bib' extending to bold black streaks on breast and flanks (all pale-fringed in winter); back boldly streaked black, merging at sides with black of breast.* Female not always safely separable, though supercilium and underparts are whiter, with breast and flanks grey-streaked. Often gathers in large, compact flocks on passage/winter. **Voice** Song as House Sparrow, but faster and audible at long range; calls slightly higher than in House Sparrow. **Habitat** Rural settings, open cultivation in winter. **Note** Has bred Bahrain, Qatar; passage and winter hatched, rare in SE Arabia.

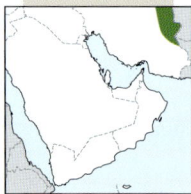

Eurasian Tree Sparrow *Passer montanus* V
L: 14. Smaller, more delicate than House Sparrow with *chestnut crown, black spot on whiter cheeks and a smaller black 'bib'; two narrow white wing-bars* and grey-brown rump. **Voice** Song is a series of repeated *tweet* notes. Calls clearer and harder than House Sparrow, and higher pitched. Flight call is a hard *tek-tek-tek*; also *tchu-wit, pilp* and a hard *chik*. **Habitat** Parks, gardens or woodland, often near habitation. **Note** Winter hatched; vagrant Oman, UAE.

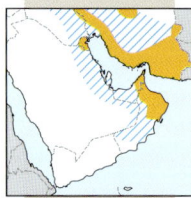

Yellow-throated Sparrow *Gymnoris xanthocollis* PM, mb
L: 13.5. Olive grey-brown unstreaked sparrow, with long, stout-based, pointed bill – black in breeding male, pinkish-brown in female and non-breeding male. Male has yellow spot on lower throat, chestnut lesser coverts and broad white wing-bar. Female and juvenile lack chestnut lesser coverts and yellow throat-patch but have distinctive bill shape and prominent wing-bar; grey legs, pointed bill and all-dark tail separates from Pale Rockfinch. Perches in trees; finch-like movements on ground; dipping flight rather pipit-like. May migrate by day in flocks. **Voice** Quiet chirruping song softer, more melodious and rhythmic than House Sparrow. Call a sparrow-like *cheep, chilp* or *chirrup*. **Habitat** Open dry woodland, date groves, cultivated areas. Nests in hole or crevice in tree. **Note** Passage hatched; rare in winter Oman, UAE. [Alt: Chestnut-shouldered Petronia]

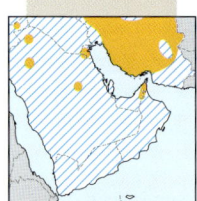

Pale Rockfinch *Carpospiza brachydactyla* PM, mb
L: 15. *Unstreaked, grey-brown, lark-like sparrow with whitish wing-panel and wing-bars, long primary projection, short dark tail with white tip* (obvious in flight), strong pale bill (with curved culmen) and prominent dark eye in pale face. Sexes similar. In flight has a lark-like appearance. Often gregarious outside breeding season. Feeds in manner of lark, otherwise hops almost upright. **Voice** Distinctive insect-like buzzing song monotonous and persistent *tss tss tss tseeeeeeeeei*. Flight call a soft trill, recalling distant European Bee-eater; also *piyee* or *twee-ou*. **Habitat** Rocky and scrubby areas at low to moderate altitudes; cultivation on passage. Untidy, domed twig and grass nest built in low bush. **Note** Passage hatched; some winter in SW Arabia. [Alt: Pale Rock Sparrow]

PLATE 106: WEAVERS, BISHOPS AND QUELEA

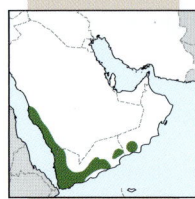

Rüppell's Weaver *Ploceus galbula* RB
L: 14.5. *Male's golden-yellow plumage, streaked mantle, dark chestnut mask with black surround to bill diagnostic*, in non-breeding season loses mask. *Female olive-brown above with dark streaks, buffish-white below washed yellow-buff on throat and breast.* Gregarious, often forming large flocks; noisy. **Voice** Song a wheezy chatter ending in insect-like hissing sounds; call is a dry *cheee-cheee*. **Habitat** Crops, palm groves, savanna, wadis with acacias and other bushes up to 2500m, commonest at lower altitudes. Nest suspended from acacia or other tree; colonial. Parasitised by Diederik Cuckoo. **Note** Only native weaver in region.

Streaked Weaver *Ploceus manyar* E/I
L: 14. Short-tailed and thickset with heavy bill. Male with *yellow crown and prominent streaks on breast*. Female and non-breeding male brown, *streaked on breast* and upperparts, including crown; *dark submoustachial streak* and yellowish supercilium. **Voice** Song unmusical strained jumble; calls loud *chirt*. **Habitat** Wetland scrub and reeds. **Note** Non-native escape; breeds Kuwait, Oman, Qatar, Saudi Arabia, UAE. Established at Al Ansab Wetland near Muscat.

Yellow-crowned Bishop *Euplectes afer* E/I
L: 10. Small and compact; breeding *male black below, with black hindneck collar and sulphur-yellow crown, back and rump*. Non-breeding male, female and juvenile streaked on sides below with yellowish supercilium. **Voice** Buzzing and chipping notes. **Habitat** Grassland, marshes, scrub. **Note** Non-native escape; recorded Oman, UAE.

Southern Red Bishop *Euplectes orix* E/I
L: 11–12. Sparrow-sized but short-tailed. Breeding male fiery red above with *black on forehead and throat; breast-band red, belly black*. Female, juvenile and non-breeding male flat-crowned; buffy and finely streaked below, bill small and pointed. (Similar Northern Red Bishop *E. franciscanus* [not illustrated] occasionally reported; *male has throat red and black on head reaches crown*. Female doubtfully separable.) **Voice** Call is thin, high, squeaky *cheet*; song consists of buzzy chirping. **Habitat** Fields, grassy scrub, waterside canes. **Note** Non-native escape; recorded Oman, UAE.

Red-billed Quelea *Quelea quelea* E/I
L: 12. Stocky and short-tailed. Male in breeding dress distinctive; non-breeding male and female *streaked above, with prominent supercilium and all-red bill*. **Voice** Chipping chatter. **Habitat** Grasslands, reeds. **Note** Non-native escape; confirmed near Muscat.

226

PLATE 107: AVADAVAT, WAXBILL, MUNIAS AND SILVERBILLS

Red Avadavat *Amandava amandava*   E/I
L: 9–10. *Male crimson with numerous white spots on underparts and wing-coverts. Female brown above, buffish below with fulvous-yellow belly; rump red, tail blackish*. Non-breeding male as female but greyer below. Coral-red bill (both sexes) with black culmen. Juvenile as female but with buff wing-bars and dark bill; lacks red rump. Often in small, low-flying flocks. **Voice** Song is high-pitched, continuous twittering; call is thin, high *teee* or *tsi*. **Habitat** Damp scrub, reeds. **Note** Non-native escape; breeds Bahrain, N Oman, Kuwait, Saudi Arabia, UAE.

Common Waxbill *Estrilda astrild*   E/I
L: 10. Adult is brown above including rump and tail. Underparts slightly paler with fine barring. *White cheeks and red line through eye. Bill bright red*. Immature duller with black bill. **Voice** Chipping contact calls. **Habitat** Cultivation, thickets. **Note** Non-native escape; breeds near Muscat.

Tricoloured Munia *Lonchura malacca*   E/I
L: 11. Adult has black head, upper breast and belly. *Lower breast white*. Upperparts rufous-brown. Bill silver-coloured. Immature is uniform brown. **Voice** Trisyllabic chirping call in flight. **Habitat** Cultivation and farmland. **Note** Non-native escape; encountered in Muscat and Salalah areas. [Alt: Black-headed Munia]

Chestnut Munia *Lonchura atricapilla*   E/I
L: 12. All *chestnut with black head, throat and neck, and blue-grey bill*. Immature similar but paler. **Voice** A weak pee, pee. **Habitat** Weedy waste ground, cultivation. **Note** Non-native escape; recorded near Muscat.

Scaly-breasted Munia *Lonchura punctulata*   E/I
L: 12. Adult chestnut or dull tan above with white *underparts brown-scaled*. Juvenile plain brown above and below. Gun metal bill in adult, duller in juvenile. Flight bouncing; often in rapidly moving flocks. Feeds on grass seed-heads. **Voice** Calls a high piping *bee-bee*, recalling Eurasian Siskin; song soft, thin whistles and slurred notes. **Habitat** Waste and grassy areas. **Note** Non-native escape; breeds Oman (Muscat and Salalah), Saudi Arabia, UAE.

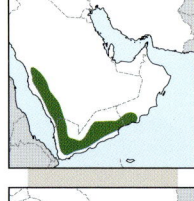

African Silverbill *Euodice cantans*   RB
L: 11. Similar to Indian Silverbill, from which geographically separated. Distinguished by *fine vermiculations on wing-coverts and tertials* (only visible at very close range), *and black rump and uppertail-coverts*; may also show slight brownish on chin. **Voice** Similar to Indian Silverbill, but higher pitched, squeakier. Song a rapidly repeated high-pitched trill comprising single, then double notes; phrases descending then rising. **Habitat** As Indian Silverbill. **Note** Native to SW Arabia; some releases outside mapped distribution.

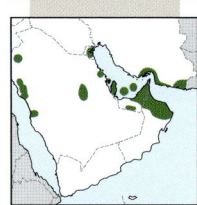

Indian Silverbill *Euodice malabarica*   RB
L: 11. Small with *large conical silver-grey bill, prominent eye and pointed black tail*. Similar to African Silverbill but told by *whitish rump and uppertail-coverts, and absence of vermiculations on wing-coverts and tertials*. Juvenile has pale edgings to wing feathers. Fairly tame and frequently in small groups, sitting close when perched; often waves and flicks tail. Flight undulating. **Voice** Rapid, tinkling *cheet cheet cheer* flight call; short high-pitched, trilling *zip-zip*; harsh *tchwit* and conversational *seesip seesip*; song a short trill. **Habitat** Hills and wadis, grassland, scrub, cultivation, palm groves, gardens. Nests in bush, crevice or old nest of sparrow; builds suspended nest. **Note** Native to SE Arabia, but western breeding populations originate from escapes.

PLATE 108: ACCENTORS AND WAGTAILS I

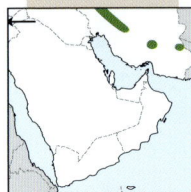

Radde's Accentor *Prunella ocularis* V

L: 15.5. Long, broad *white supercilium, blackish crown and ear-coverts, whitish throat, dark-spotted malar stripe and peach wash on unstreaked breast* (but often boldly streaked flanks) distinctive. Juvenile similar to adult but crown brownish, streaked darker; sides of throat, breast, and flanks more heavily streaked; underparts paler buff-white. **Voice** Calls and song Dunnock-like; song weaker and slower, rising and falling, with trembling quality, *di-diii-diii-diii-diii* or *slee-vit-chur-chur-tui*. **Habitat** Low scrub and boulders in mountains, 2500–3500m; in winter down to 1000m. **Note** Partial migrant; winter hatched; vagrant Kuwait, Oman.

Black-throated Accentor *Prunella atrogularis* V

L: 15. *Black throat-patch* diagnostic, though in autumn/winter less well defined or even hidden by pale fringes (mainly first-winter females); such pale-throated individuals have *supercilium palest/whitish in front of eye and not concolorous with throat and breast*. First-winter individuals told from Radde's Accentor by black throat, paler, slightly streaked brown crown and ear-coverts (not solid dark), off-white to buff supercilium, absence of dark malar, and slightly warmer brown upperparts. **Voice** Call a soft *trrt*. **Habitat** Weedy patches, scrub, gardens in mountain villages. **Note** Vagrant Kuwait, Oman.

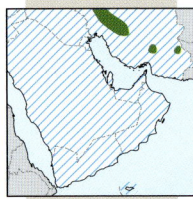

Forest Wagtail *Dendronanthus indicus* wv

L: 17. Unique in having *two broad blackish breast-bands, one across upper breast, the other below, broken*; underparts otherwise white, washed very pale yellow or creamy. Upperparts olive-brown, wings dark with *broad buff-white wing-ba*rs; also a thin white supercilium; tail blackish with white outer feathers. Pipit-like movements on the ground, *usually in woodland; does not wag tail but sways body from side to side*. Perches freely in under-canopy, where it remains when disturbed from ground. **Voice** Call a loud, abrupt Chaffinch-like *pink* or *pink-pink*. **Habitat** On ground in clearings in damp forest or plantation, often near water. **Note** Very rare passage and winter visitor; vagrant Kuwait.

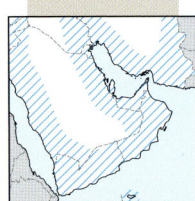

Grey Wagtail *Motacilla cinerea* wv

L: 18. In all plumages has *yellow vent and undertail-coverts, greyish back, bold white translucent bar at base of flight feathers (in flight), very long tail and extremely undulating flight*. Male in summer *has black throat and white sub-moustachial stripe*; female has less black or even whitish throat; immature has white throat and buff, not yellow breast. *Legs pinkish-brown* (black in other *Motacilla* wagtails). Tail-wagging more pronounced than in other *Motacilla* wagtails. **Voice** Call distinctive, resembles White Wagtail but harder, more metallic and high-pitched, a piercing *tzi-lit* or *tsiziss*, or loud *chink*, usually given in bouncing flight. Song a distinctive series of sharp mechanical notes. **Habitat** Wooded streams, wadis, pools, trickles. **Note** Passage and winter hatched.

Citrine Wagtail *Motacilla citreola* PM, WV

L: 18. Male unmistakable; subspecies *citreola* has *bright yellow head* and underparts, *black neck-band and grey upperparts*; *werae* (breeds Turkey) has no (or reduced) neck-band and less-rich yellow underparts; *calcarata* male in summer has *pitch black mantle* and largely white coverts. *Broad pure white double wing-bars characteristic at all ages*. Female from Western Yellow Wagtail by *greyish upperparts* (tinged olive-brown in Western Yellow), *yellow supercilium surrounding grey-brown cheeks merging on sides of neck and yellowish throat*. First-autumn birds have grey upperparts, *pale surround to ear-coverts, often pale forehead and lores, whitish underparts without yellow*, prominent wing-bars and broadly fringed dark tertials. **Voice** Call pronounced buzzy *tsreep*; sometimes a double *zielip*; some eastern '*flavas*' call rather similarly. Song recalls Western Yellow Wagtail, generally shriller. **Habitat** Breeds 1,500–2,500m in swampy meadows or near streams. Outside breeding season near fresh water, lagoons or sewage ponds. **Note** Passage and winter hatched (all races), but very rare in north in winter. Subspecies *calcarata* breeds NE Iran; vagrant Oman.

PLATE 109: WAGTAILS II

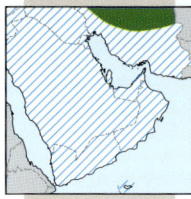

White Wagtail *Motacilla alba* WV, PM
L: 18. Race *M. a. alba* predominates. Easily told from other wagtails by *grey, black and white plumage*, with grey back contrasting with black nape (male) or crown (female), white face and ear-coverts contrasting sharply with black throat and breast. Juvenile, and especially winter birds, have whitish underparts *with prominent black sagging half-moon on breast*. Flanks show some greyness, concolorous with or paler than mantle. Flight undulating, when often calls. **Voice** Disyllabic call slightly metallic, a hard *tse-lit* with stress on second syllable. Song composed of call notes, a lively garbled twittering. **Habitat** Open areas, often near habitation, cultivation, livestock. **Note** Passage and winter hatched. Two other distinct subspecies are described separately below.

Masked Wagtail *M. a. personata* NE Iranian breeding subspecies has *black of head and breast merging at sides of neck and ear-coverts, leaving smaller white area around face* (same pattern in both sexes, but darker in male); white edges on wing-coverts, tertials and secondaries also much broader than in White Wagtail. **Voice** Call more metallic than White Wagtail, approaching Grey Wagtail. **Habitat** Open areas, often near habitation, cultivation and especially near cattle. **Note** Winter range hatched.

Amur Wagtail *M. a. leucopsis* Adult male in summer plumage is unmistakable and differs from White Wagtail (nominate race) by larger area of white on face, *black back* (grey in White Wagtail), *white flanks and a large white wing panel*. **Voice** Similar to White Wagtail. **Habitat** In winter, mainly farmland. **Note** Vagrant Oman.

Eastern Yellow Wagtail *Motacilla tschutschensis* V
L: 17. Race *M. t. taivana* (Green-headed Wagtail) occurs. Spring male has *crown, nape and back greyish-green, yellow supercilium and underparts*, white wing-bars and *black ear-coverts* that are noticeably darker than forehead, crown and nape. Note that black ear-coverts are not surrounded by a pale ring as in immature Citrine Wagtail; some birds occurring in winter in the Salalah area resemble *taivana* but show yellow at the rear edge of the dark ear-coverts. The origin of these birds is unclear but it has been suggested they could be of hybrid origin. **Voice** High-pitched *zii*. **Habitat** Wet meadows, grasslands. **Note** Vagrant Oman.

PLATE 110: YELLOW WAGTAILS

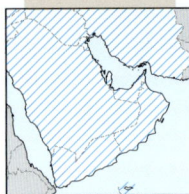

Western Yellow Wagtail *Motacilla flava*   PM, WV
L: 16.5. This species complex has many subspecies. In spring and summer the head colour of males varies as shown below; hybrids occur, making some identifications uncertain. Females are all similar: **yellowish below with whitish throat**, and most are unidentifiable to subspecies. **Voice** Calls: *flava* high *see-u*; *thunbergi* slightly sharper; *beema* more buzzy; *feldegg*, see below. **Habitat** Wet or dry grassland, open fields. **Note** Passage hatched for all subspecies; generally scarce in region in winter. Seven subspecies have been identified in Oman, and these are described separately below (roughly in order of abundance in Oman).

Sykes's Wagtail *M. f. beema* Pale greyish crown and ear-coverts, long, broad white supercilium and whitish chin. Some first-autumn birds lack yellow below (rear of underparts normally tinged yellow in immatures of other subspecies), are greyish above (normally tinged olive-brown in other subspecies) and show fairly prominent whitish wing-bars, thus resembling first-autumn Citrine Wagtail (call also rather similar). **Note** Passage throughout; some winter in S Arabia (including Oman).

Black-headed Wagtail *M. f. feldegg* Male has head and nape glossy black, sharp demarcation to dark mossy-green mantle, entire underparts bright yellow; duller in winter. Female superficially like male, often almost monochrome, being dark-headed with greyish upperparts and only yellow-sullied underparts. **Voice** Song loud, simple, *sree-srriep*, stress on second syllable. Call strong ringing *psreee-u*. **Note** Abundant in Oman; has bred successfully. Includes '*melanogrisea*' which resembles *feldegg* but back is paler olive, underparts paler yellow, chin and sides of throat white. Breeds E Iran; passage Arabia (including Oman).

Grey-headed Wagtail *M. f. thunbergi* Dark grey crown, with no sharp demarcation to upperparts but contrasting with darker ear-coverts, usually no supercilium or white on chin and throat. **Note** Widespread on passage, abundant Oman, UAE; some winter in Arabia.

Yellow-headed Wagtail *M. f. lutea* Whole head yellow, some with olive crown and ear-coverts, upperparts yellow-olive. **Note** Passage in Near East and Arabia; scarce Oman, UAE.

Blue-headed Wagtail *M. f. flava* Blue-grey crown and ear-coverts; long white supercilium; sides of throat sometimes white. **Note** Passage mainly in western Arabia; few winter; irregular Oman.

Ashy-headed Wagtail *M. f. cinereocapilla* Recalls *thunbergi* but throat whitish. **Note** Scarce on passage in Near East; rare or vagrant Saudi Arabia and Oman.

White-headed Wagtail *M. f. leucocephala* White head, tinged blue or grey on rear-crown and ear-coverts. Broad pale yellow wing-bars, moss-green mantle, tertials broadly edged pale yellow, 'thighs' white-feathered. **Note** Vagrant NE Iran, Kuwait, Oman, UAE.

Hybrids
'*superciliaris*' (*feldegg* x *flava* or *beema*) Dark slaty-blue head with clean white supercilium and submoustachial, broad yellowish wing-bars. **Note** Passage Near East and Arabia.

'*dombrowskii*' (*feldegg* x *flava* or *beema*) Similar to *flava*, but with blackish ear-coverts. **Note** Scarce on passage in Near East and Arabia.

'*xanthophrys*' (*feldegg* x *lutea*) Supercilium yellow, crown, lores and ear-coverts contrasting solidly dark, underparts yellow (not illustrated). **Note** Rare on passage or winter in Near East and Gulf States.

PLATE 111: PIPITS I

Golden Pipit *Tmetothylacus tenellus*   V
L: 15. Male with *vivid yellow wings and tail-sides, obvious in flight, sulphur-yellow underparts with black band across breast*. Female with primaries edged yellow, underparts buffy with belly yellowish; immature similar with streaked breast. Perches freely on trees. **Habitat** Open dry bush country. **Note** Vagrant Oman and Yemen, from Africa.

Richard's Pipit *Anthus richardi*   WV, PM
L: 18. Large, robust pipit; adult from smaller, sandier Tawny Pipit *by longer tail and legs, stouter bill, typically more upright stance, prominently dark-streaked brown upperparts, streaked breast and characteristic call*. From first-autumn Tawny Pipit, some still with well-streaked breast, *by pale lores* (dark in Tawny, bleaching whitish). More undulating flight and longer tail give more wagtail-like flight than Tawny; *frequently hovers just before landing* (Tawny rarely does). **Voice** Flight call loud, harsh throaty *schreeip*; also less distinctive Tawny-like *tjiirrup*. **Habitat** Grassland, fields, marshes. **Note** Passage hatched but often rare; winters in SE Arabia.

Blyth's Pipit *Anthus godlewskii*   V
L: 17. *Very like Richard's Pipit but slightly smaller, shorter-tailed (noticeable in flight) and shorter-legged*, with slightly shorter, deep-based but pointed bill (Richard's culmen downcurved at tip); supercilium often shorter. Outer tail feathers also differ (see plate). Crown neatly streaked, lores pale. Mantle and breast streaked, on latter often a neat fine gorget; flanks sometimes buffy. Hindclaw of medium length (long in Richard's). Median wing-coverts of adult (if not worn), and first-autumn birds with some renewed to adult type (but not otherwise), have *dark centres sharper, more squarely cut-off against pale tips than in Richard's, producing a prominent pale bar* (in Richard's, dark centres protrude centrally, with pale edging more diffuse). Juvenile more streaked above and on breast than adult, with whiter fringes and tips to wing-coverts more clear-cut. First-autumn birds from young Tawny Pipit by pale lores, more streaked upperparts, warmer, browner plumage, distinctly streaked breast, and narrower, more clear-cut pale tertial fringes. Drops into grass without hovering. **Voice** High-pitched *shreuu* or *spzeeoo*, unlike Richard's flight call, closer to slightly hoarse *sweeuu* call of flava wagtails; also a short *chep, chip, chup* or *pip* reminiscent of Tawny Pipit; sometimes joined into unique *schreuu-chup chup*. **Habitat** Fodder fields, other grassy areas and open ground. **Note** Passage and winter hatched; vagrant Bahrain, Oman.

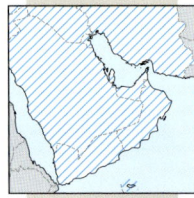

Tawny Pipit *Anthus campestris*   PM, WV
L: 16.5. Slim, largish, upright pipit with *relatively long tail, legs and bill*. Adult has *poorly streaked sandy upperparts, nearly unstreaked breast*, plain sandy wings with conspicuous dark-centred median coverts, and bold whitish supercilium. First-autumn birds, still with streaked breast, *lack bold flank-streaks of most smaller pipits*. Legs pinkish-salmon. Flight undulating; runs quickly, stopping suddenly. **Voice** Typical flight call is sparrow-like *chilp* or *chirrup*. Song, usually in undulating songflight, is simple, thin and metallic, *zriiliu, zseer-lee* or *ziu-ziirliu*. **Habitat** Sparsely vegetated ground, cultivation, plains; any open country on passage. **Note** Passage hatched; winters S Iran, and most of Arabia.

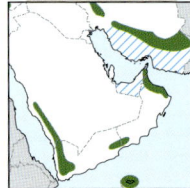

Long-billed Pipit *Anthus similis*   RB, wv
L: 17. Large; from Tawny Pipit by *creamier-buff flanks and vent*, less pronounced malar streak and *creamy-buff outer tail feathers* (which bleach whitish to resemble Tawny). *Legs pinkish-yellow*. Bill longer with more drooping tip than Tawny, with at least culmen dark. *Upperparts grey-brown, tinged olive in some*, pale edges and dark centres to wing-coverts and tertials generally less pronounced than Tawny, supercilium often narrower and ear-coverts plainer brown. *Tail looks long, broad and dark* in flight; *when perched tail often flicked upwards and fanned outwards*. Seldom well-streaked on breast and upperparts. From Richard's by plainer head pattern, less upright stance, *less streaked plumage and buff outer tail feathers* (always white in Richard's). **Voice** Flight call loud, clipped *tjuip* or *che-vee*; also rich *tchup* and quiet, soft *tchut*. Rising and falling song given in undulating songflight, *duiit-diuuu, peet-trueet* or *shreep chew-ee*. **Habitat** Rocky hills, mountain slopes with scattered vegetation, up to 3,000m. **Note** Partial and altitudinal migrant; winter hatched.

PLATE 112: PIPITS II

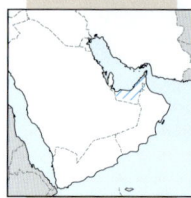

Buff-bellied Pipit *Anthus rubescens*   V
L: 15.5. Subspecies *japonica* resembles dark winter Water Pipit; *upperparts dark olive or grey-brown, faintly streaked; breast and flanks usually rather boldly streaked* (in Water Pipit, sparser, finer, particularly on flanks); *legs pale brown or reddish-brown* (normally darker in Water Pipit); *white eye-ring more obvious, malar streak often broader, ending in black patch at sides of neck*. In spring, underparts pinkish-buff, breast lightly spotted, *but flanks still strongly streaked*; upperparts almost unstreaked olive-grey. Larger and darker than young Red-throated with faintly streaked upperparts. **Voice** High-pitched, short *tripp*, lacking shrill quality of Water Pipit but similar to Meadow Pipit. **Habitat** Grassy fields, near water. **Note** Passage and winter hatched, but rare; vagrant Iran, Kuwait, Oman, Qatar.

Water Pipit *Anthus spinoletta*   WV
L: 16. Largish with *dark brown or red-brown legs* and greyish indistinctly streaked mantle. In winter browner above, with *whitish underparts sparsely streaked, almost unstreaked in some*. In summer *breast variably tinged rosy-pink and practically unstreaked*; creamy-white supercilium generally pronounced. **Voice** Call sharpish ringing *tsrieh* or *bzisp*. **Habitat** On passage and in winter, in lowland grassland, wetlands, lakesides. **Note** Passage and winter hatched, often abundant.

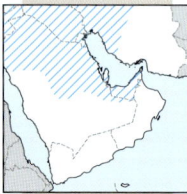

Meadow Pipit *Anthus pratensis*   wv
L: 14.5. Similar to Tree Pipit but more slightly built, with *long hindclaw;* from first-autumn Red-throated Pipit by *almost unstreaked rump, more broken breast- and flank-streaks, call* and less contrasting mantle pattern. Flies with irregular undulations and changes of direction; on ground flicks tail nervously. **Voice** Flight call thin, nervous, *sit–sii–sit* or *tsis–sip*. **Habitat** Open county, marshes, coasts. **Note** Passage and winter hatched; rare Oman.

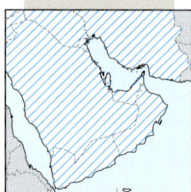

Tree Pipit *Anthus trivialis*   PM, wv
L: 15. Slightly stockier than Meadow Pipit with slightly shorter tail and deeper-based bill. Warmer, less olive, above, warmer buff on breast and whiter belly; breast often boldly striped, but flanks more finely streaked than Meadow; dark malar and creamy submoustachial streak generally bolder in Tree Pipit but moustachial streak fainter. *Pale hindclaw short, legs pinkish-salmon* (dirty flesh to orange-brown in Meadow). In autumn, from Red-throated Pipit *by voice, unstreaked rump, fine flank-streaks*, less variegated streaking above and smaller blackish spot at end of malar. Bounding flight more direct than Meadow Pipit; on ground pumps tail, unlike Meadow's nervous flicking. **Voice** Flight call distinctive, short *bzeez* or *speez*; alarm a repeated *stit*. Contact note distinctive quiet *tip*. **Habitat** Open woodland, parkland and grasslands. **Note** Passage hatched; rare in south in winter.

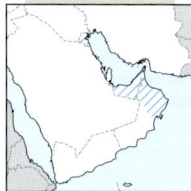

Olive-backed Pipit *Anthus hodgsoni*   wv, pm
L: 14.5. Resembles Tree Pipit, *with broad supercilium, white behind eye, buff in front, edged black above, pronounced small white and black spots on rear of ear-coverts* (only sometimes weakly present in Tree); *upperparts greener, in subspecies yunnanensis vaguely or diffusely streaked* unless worn (Tree Pipit clearly streaked); *breast more boldly streaked black*, some also show *bold blackish flank-streaks*. Subspecies *hodgsoni* similar but mantle obviously streaked (extreme vagrant to UAE). Behaviour as Tree Pipit, though tail-wagging usually more pronounced. **Voice** Flight call a thin *tzeez*, usually drawn-out, sometimes slightly stronger at start. **Habitat** As Tree Pipit. **Note** Winter hatched, but rare; vagrant Bahrain, Iran, Saudi Arabia.

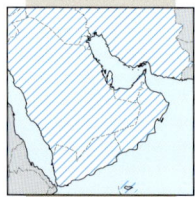

Red-throated Pipit *Anthus cervinus*   PM, WV
L: 15. Adult has variable amount of *pinkish or reddish-buff on face, supercilium, throat and upper breast*; retained in autumn though rarely in winter; some females have only beige throat. From Meadow Pipit in autumn/winter *by call; boldly streaked rump*, breast and flanks creamy or white (usually pale buff in Meadow), *flanks usually with 2–3 bold, almost unbroken, stripes; malar streak ends in large, dark triangle on side of throat*; upperparts with *bolder blackish and creamy white stripes*. Flight similar to Meadow Pipit. **Voice** Distinctive, thin, high-pitched, drawn-out *pseeee*, slowly dying away. **Habitat** Marshes, grassland, cultivation; usually near water. **Note** Passage and winter hatched; rare in north in winter.

PLATE 113: FINCHES

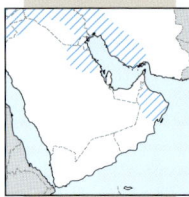

Brambling *Fringilla montifringilla* wv
L: 15. In any plumage told by *rusty-orange breast, shoulder and wing-bar, white belly, mottled upperparts and, in flight, white rump*. In winter (when only likely to be seen in the region) *shows pale patch on nape and yellowish bill with dark tip*. Often in flocks; mixes freely with buntings in winter feeding areas. Flight erratic. **Voice** Distinctive, in flight or perched, a loud, unmusical nasal *jehp*, or softer repeated *tip* or *pip*. **Habitat** Woodland, fields or other cultivation, often amongst or under trees. **Note** Winter hatched; vagrant Bahrain, UAE.

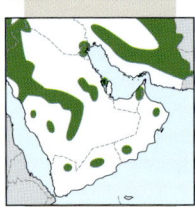

Trumpeter Finch *Bucanetes githagineus* rb
L: 14. Small, ground-dwelling finch with large head, stout bill and rather short tail. Male distinctive with *grey head, pinkish wash on forehead, underparts, rump and wings and orange-red bill*. Non-breeding male, female and juvenile/first-winter are plain sandy grey-buff with slightly paler rump, blackish-grey wings and tail with paler feather-edgings, and pale yellowish-brown bill; some breeding females develop faint pink wash on plumage; *legs in all plumages orangey-flesh*. **Voice** Song distinctive, drawn-out nasal, wheezing buzz *cheeeee*; call short *chee* or *chit*; in flight a soft *weechp*, most calls with buzzing quality. **Habitat** Bare rocky and stony hillsides and wadis, stony desert; visits pools and waterholes. **Note** Uncommon breeding resident in Oman. Nomadic with irregular post-breeding wanderings.

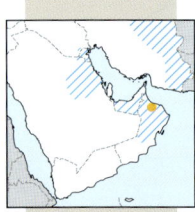

Common Rosefinch *Carpodacus erythrinus* pm, mb
L: 14. Compact finch with *stout bullfinch-like bill* and round head. Adult male easily told by *red head, breast and rump contrasting with brown upperpart*s. Female, first-summer male (which often sings) and juvenile are dull olive-brown, lightly streaked above and *more heavily streaked below, with two whitish or buffish wing-bars*, without wing-panel in flight feathers; uniform head shows conspicuous dark beady eye. Often sings from prominent position, otherwise rather inconspicuous. **Voice** Song diagnostic, a clear, lively loud whistle *vii-dji vii-di-djiv-viuuu* (rendered as 'pleased to meet youuu'). **Habitat** On passage anywhere with trees or scrub. **Note** Passage hatched; rare in winter in Arabia; has bred N Oman near Jabal Shams.

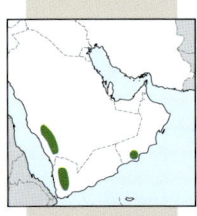

Yemen Serin *Crithagra menachensis* rb
L: 11.5. Small, rather plain grey-brown finch with lightly streaked underparts. Note *crown-streaks, indistinct moustachial stripe in front of pale cheek-patch, and rather small bill*. Largely ground-dwelling, often seen on buildings or clinging to walls. Frequently occurs in small flocks. **Voice** Most characteristic call is a hard *che-che-che*; also twittering *chirrip, chirrup*; quick repeated *prlyit-prlyit* and wagtail-like *cheir-virp cheir-virp*. Feeding flocks have musical, whispering *tleet-tleet*. Song is a rippling *chew-chee-chee-chew*, either in dipping songflight or perched. **Habitat** High plateaux and rocky hillsides, at 1000–3200m; also in villages; trees are not essential; nests in hole in rock or wall. **Note** Highly localised range in S Oman.

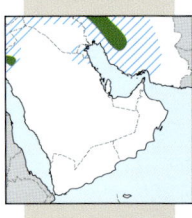

European Goldfinch *Carduelis carduelis* v
L: 14. Colourful and *readily identified by bold red, black and white head pattern, striking broad yellow bar on black wings, particularly obvious in flight*, when white rump also conspicuous. Juvenile has pale greyish body plumage and unmarked head; identified by yellow wing-bar, white rump and call. **Voice** Call, often uttered in flight, is a characteristic, repeated, liquid *tick-le-lit* with stress on last note; rasping notes sometimes heard from flocks. Song a liquid twitter, with calls mixed with song. **Habitat** Scrub, woodland, orchards, cultivated areas; open country, weedy wasteground. **Note** Passage and winter hatched; vagrant Bahrain, Oman, UAE. A popular cagebird, with escapes frequent.

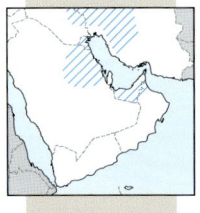

Eurasian Siskin *Spinus spinus* v
L: 12. Small, finely built finch; *male yellowish-green with black forehead and chin, broad yellow wing-bar, sides of tail and rump, particularly obvious in flight*. Female (which lacks black on head) is olive-brown above, white below but has tinge of yellow-green, which, together with yellowish in wings, tail and on rump, aid identification. Back and flanks streaked dark in all ages. Often feeds hanging upside-down in cone-bearing trees. Appears small and short-tailed in undulating flight, often in tight flocks. **Voice** Call a drawn-out, high-pitched *tsee-u*; from feeding flocks a fast, dry *kettkett* and twittering notes. **Habitat** Parks, plantations, *Casuarina* trees. **Note** Passage and winter hatched, sporadic and rare in south; vagrant Bahrain, Oman, Qatar.

240

Brambling

Trumpeter Finch

Common Rosefinch

Yemen Serin

European Goldfinch

Eurasian Siskin

PLATE 114: GROSBEAK AND BUNTINGS I

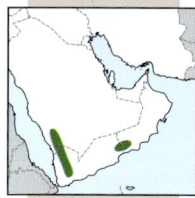

Arabian Golden-winged Grosbeak *Rhynchostruthus percivali* rb
L: 15. Plump, sparrow-sized finch with *very stout blackish bill*. Easily identified by *white cheeks* and *yellow patches in wings and tail*. Female differs from male in having duller plumage without blackish area around base of bill. Juvenile has brownish head and prominently streaked upper- and underparts. Often in loose groups and can be inactive, sitting unobtrusively amongst branches. Bounding flight, often flies some distance if disturbed. **Voice** Liquid, discordant song often starts with *whit-whee-oo* or *tvit-te-vyt-te-vict*, repeated and interspersed with European Goldfinch-like notes, often in fluttering, bat-like, display flight. Varied calls include *wink*, and soft, Goldfinch-like, *tlyit-tlyit*. **Habitat** Hillsides and wadis with euphorbias, acacias and other fruit-bearing trees, 250–2000m. **Note** Mostly a scarce resident in Dhofar mountains, but moves according to food availability.

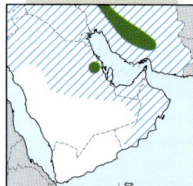

Corn Bunting *Emberiza calandra* pm, wv
L: 18. Bulky with fairly large head, *large conical bill and no white in tail*. Underparts heavily streaked, often merging into irregular black patch on breast; often pale submoustachial stripe and variable ill-defined malar streak. *Heavy, rather fluttering flight, legs often dangling*; lacks white on trailing edge of wings shown by Skylark, with which it can be confused in flight. Often perches on telegraph wires or bush-top; flocks outside breeding season. **Voice** Song, often from exposed perch, accelerates, ending with monotonous almost insect-like sound *tick-tick-tzek- zee-zri-zizizizi*. Calls single hard, almost clicking *twik* or *tritt*, sometimes quickly strung together in series. **Habitat** Open farmland, irrigated grasslands, bushy areas. **Note** Passage and winter hatched.

Pine Bunting *Emberiza leucocephala* V
L: 16.5. Similar to Ortolan Bunting in build and size. *Male's black-bordered white patch on crown and cheeks in otherwise chestnut head and throat diagnostic* (subdued in winter). Grey-brown female has streaked crown, mantle, breast and flanks, and chestnut rump; *some show a little whitish on crown and chestnut on whitish throat*. Many females, and all first-autumn birds, resemble Yellowhammer (illustrated for comparison), *but yellow in plumage replaced by white, including belly and fringes to primaries*; lesser coverts more uniform grey-brown and bill more frequently bicoloured than Yellowhammer (dark grey upper, pale grey lower mandible). **Voice** Metallic, sharp *staeup*; also a nervous *trr-rrr-rrr-ick*. **Habitat** Farmland, open country with bushes. **Note** Vagrant Oman, Saudi Arabia, UAE.

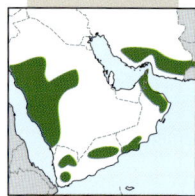

Striolated Bunting *Emberiza striolata* RB
L: 14. *Small rufous-coloured bunting with orange-yellow lower mandible (adult) and rufous-edged outer tail feathers; streaks on upperparts and wing-coverts thin and vague. Typical male has head striped dark ash-grey and white*, in some individuals poorly defined; *throat and upper breast are steel-grey with black speckles; upperparts, wing-coverts, lower breast and belly bright rufous*. Female duller with more diffuse head pattern. Often fairly shy. **Voice** Song *wi-di-dji-du-wi-di-dii* or *witch witch a wee*. Calls include squeaky *tzswee*, nasal *dwiib*; *dweek* and *sweee-doo*. **Habitat** Oases, desolate rocky wadis and hills with little vegetation. Nests in hole in building, wall or rock crevice. **Note** Common resident in N Oman; less common in S Oman. Vagrant Bahrain, Qatar.

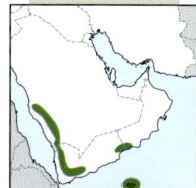

Cinnamon-breasted Bunting *Emberiza tahapisi* RB
L: 16.5. Medium-sized *with orange-yellow lower mandible, bold black streaks on upperparts and rufous-edged outer tail feathers. Male has black throat and four white stripes over black head*. Female duller with less pure black-and-white head pattern. Juvenile resembles female but head and throat grey-brown, latter flecked blackish onto upper breast; pale head-stripes buffier. Best separated from slightly smaller Striolated Bunting *by unstreaked dark throat, boldly streaked upperparts and absence of rufous on mantle and wing-coverts*. Often feeds in flocks. **Voice** Song fast, short simple jingle *dzit-dzit-dzi-re-ra* or *tru-tri-tre-ririr*, second and last note higher in pitch; song sometimes followed by scratchy notes. Calls soft metallic *anh* and a nasal *daar*; alarm a short metallic *ptik*. **Habitat** Rocky hillsides with scattered vegetation, 300–2500m. **Note** Common breeding resident in Dhofar mountains, S Oman. [Alt: African Rock Bunting]

PLATE 115: BUNTINGS II

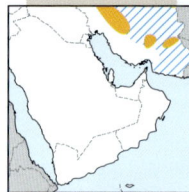

Grey-necked Bunting *Emberiza buchanani*   V
L: 16. Resembles Cretzschmar's and Ortolan Buntings in having pink bill and whitish eye-ring. *Adult has grey head, less bright than Cretzschmar's and lacks grey breast-band; underparts pale rufous* with whitish submoustachial streak and pale yellow-buff or whitish vent and undertail-coverts. In autumn (fresh plumage) underparts fringed greyish-white; *upperparts brownish-grey, indistinctly streaked, contrasting with chestnut scapulars; tertials have ill-defined dusky-brown edges without clear-cut notched pattern of Cretzschmar's and Ortolan*. Juvenile from Cretzschmar's by brownish-grey, not rufous-tinged rump, paler vent and undertail-coverts; with much less boldly streaked upperparts, paler flight feathers with ill-defined tertial pattern. **Voice** Song rich *di-di-dew, de-dew,* penultimate note higher pitched and more stressed; delivery fast or slurred. Flight call *tip, tsip* or *sik;* also *chep* or *tcheup*. **Habitat** Barren rocky hillsides and sparsely vegetated plateaux, mainly above 2,000m. **Note** Passage hatched; vagrant Kuwait, Oman.

Cretzschmar's Bunting *Emberiza caesia*   V
L: 16.5. Resembles Ortolan Bunting. *Male has bright blue-grey head and breast-band framing rusty-orange throat; rump rufous, vent and undertail-coverts rufous-buff,* blackish-brown tertials have clear-cut *rufous notched fringe*. Female duller, often with traces of grey on head and some grey on streaked breast; *from female Ortolan by rusty throat* (sometimes yellowish-buff) and warmer rump, vent and undertail-coverts. First-autumn birds resemble Ortolan *but yellow-buff throat often tinged rufous (adult plumage develops quickly)*. **Voice** Calls Ortolan-like, but sharper, *blep;* metallic *tlik* or *tlev*. **Habitat** On passage in cultivation, semi-desert. **Note** Passage hatched; vagrant Iran, Oman, Qatar, Yemen, UAE.

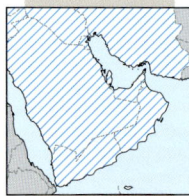

Ortolan Bunting *Emberiza hortulana*   PM
L: 16.5. Male told from similar Cretzschmar's Bunting by *olive-grey head and breast-band framing pale yellowish throat, pink bill, boldly streaked mantle and blackish-brown tertials with clear-cut light chestnut (adult) or creamy-buff notched edges*. Female duller and more streaked; separable from female Cretzschmar's by pale yellow-buff throat, brownish, not rufous-tinged rump and pale buff vent and undertail-coverts. First-autumn birds more streaked above and on breast than female; sometimes separable from similar Cretzschmar's by colour of rump, vent and undertail-coverts. Migrates in flocks. **Voice** Call a hollow, soft *plet* or *büb* and a slightly falling *sliie*. **Habitat** Open country, bare cultivation with trees and scrub, semi-desert, oases. **Note** Passage hatched.

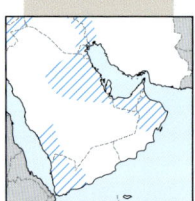

Cinereous Bunting *Emberiza cineracea*   pm
L: 16.5. Male of eastern subspecies *semenowi* (Eastern Cinereous Bunting) has yellowish underparts, including belly and undertail-coverts, upperparts olivey. Olive grey-brown female has *throat yellowish and streaked crown yellowish olive-brown;* underparts dusky-streaked on breast with belly off-white **Voice** Call short, metallic *kjip* or *djib*. **Habitat** Dry rocky slopes with sparse vegetation; also semi-deserts or bushy wadis on passage. **Note** Eastern Cinereous breeds SE Turkey, Iraq and Iran, and occurs on passage in Arabia, including Oman.

Little Bunting *Emberiza pusilla*   pm, wv
L: 13. Small and secretive. *Adult told by reddish-chestnut crown, supercilium and ear-coverts, framed by black lateral crown-stripe and ear-coverts surround narrow whitish eye-ring; white underparts with narrow black streaks on breast;* whitish wing-bars usually obvious. First-autumn told from Common Reed Bunting by *uniform pale rusty cheeks, bill shape (straight culmen), whitish eye-ring, black malar streak barely reaching bill, grey-brown lesser coverts, better defined blacker streaks on whiter underparts and call*. Separated from Rustic Bunting by *absence of rufous streaks below and grey-brown rump* (rusty in Rustic). **Voice** Call a single, hard, metallic *tik,* like Rustic Bunting but slightly sharper. **Habitat** On passage, in damp bushy and grassy areas. **Note** Vagrant to most countries of the Middle East.

PLATE 116: BUNTINGS III

Rustic Bunting *Emberiza rustica*   V
L: 14.5. Fairly robust bunting, streaked rufous below and with longish bill with straight culmen. Male unmistakable with black head broken by bold white supercilium and nape spot, bright rusty nape merging on sides of neck with rusty-red breast-band; flanks with extensive rusty streaks; rump rusty-red. Female has less blackish head with paler median crown-streak and ear-coverts. First-autumn birds confusable with Common Reed Bunting but note rufous streaks on breast and flanks, rufous rump, bill colour and shape, bolder supercilium behind eye, whiter wing-bars and call. **Voice** Call a short, distinct *zib* or *zik*, less sharp than Little Bunting. **Habitat** On passage, generally in damp, bushy or wooded areas. **Note** Vagrant to most countries of the Middle East.

Yellow-breasted Bunting *Emberiza aureola*   pm, wv
L: 15. Medium-sized with pale, stout, conical bill and white in tail-sides. Non-breeding male, female and first-autumn birds told *by prominent creamy supercilium, faint pale central forecrown-stripe with dark brown lateral crown-stripes* (less well-defined in first-autumn), *pale cheeks strongly dark-bordered, cream mantle braces and fairly distinct white wing-bars*. Rump grey-brown and heavily streaked. Underparts yellowish-white or at least sullied buffy (mainly first-autumn) with whitish undertail-coverts; flanks and sometimes breast faintly streaked. **Voice** Two main calls heard on passage, a short *tik*, not unlike that of Little Bunting, and a softer, metallic *tsiu* or *tsip*. **Habitat** Open country with trees or bushes; grassland, cultivation. **Note** Rare and declining passage migrant Oman; vagrant Bahrain, Iran, Saudi Arabia, UAE.

Black-headed Bunting *Emberiza melanocephala*   PM
L: 16.5. Heavy-bodied, stout-billed bunting *without white in tail*. Male has *black head, yellow underparts and unstreaked chestnut upperparts*; black of head subdued by pale fringes in autumn. Female and first-autumn birds lack black, bright chestnut and yellow, then appearing almost identical to Red-headed Bunting, *although upperparts usually warmer-toned* (less olive grey-brown) and underparts generally tinged more yellow, especially undertail-coverts. Often in groups on passage, particularly in spring. **Voice** Song typically bunting-like, starting with short harsh notes, ending with more ringing *tsi-tia-tia-tia-terlu-terlu-terlu*. Call a sparrow-like or Tawny Pipit-like *tjilp*; also a metallic *tlev*. **Habitat** Bushy and grassy country, open farmland. **Note** Passage hatched.

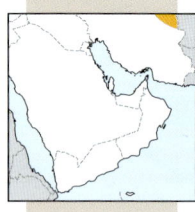

Red-headed Bunting *Emberiza bruniceps*   V
L: 16.5. Size of Black-headed Bunting, also lacking *white in outer tail feathers*. Male with *red-brown head and breast* (subdued by yellow and white in autumn), *with rest of underparts and rump yellow*. Female and first-autumn birds very like Black-headed and some not identifiable, but note *Red-headed has grey-brown, often olive-tinged upperparts including rump and scapulars* (tinged rufous in many Black-headed Buntings). Head of Red-headed Bunting slightly more uniform with less streaked forecrown and paler ear-coverts than Black-headed Bunting; underparts, particularly undertail-coverts, tend to be tinged less yellowish than Black-headed, some being uniform buffish-grey below. **Voice** Song like Black-headed Bunting. Call *bzisf*. **Habitat** Farmland and open country on passage. **Note** Vagrant Kuwait, Oman, Saudi Arabia, UAE. Frequent escape.

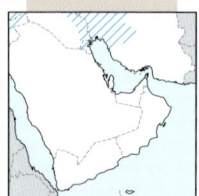

Common Reed Bunting *Emberiza schoeniclus*   V
L: 15.5. Breeding male told by *black hood and bib, white half-collar and white moustache*. Female has *broad black malar streak and whitish supercilium* on otherwise brown head. First-autumn birds resemble female but cheeks more mottled, breast and flanks more streaked. Separated from Little Bunting *by brownish (not chestnut) ear-coverts, bill with convex culmen, inconspicuous eye-ring, chestnut lesser coverts, and by call*. When perched, jerks and spreads tail nervously. In flight, undulations short and irregular, compared with the more steady flight of most other buntings. **Voice** Calls include a fine, drawn-out *tsii-u* and a metallic, slightly voiced *bzü*; in flight a low nasal *bäh*. **Habitat** Reedbeds, swampy thickets. **Note** Passage and winter hatched; vagrant Bahrain, Oman, Saudi Arabia, UAE.

REFERENCES AND FURTHER READING

OMAN
Eriksen, J. & Victor, R. 2013. *Oman Bird List (Edition 7)*. Center for Environmental Studies and Research, Sultan Qaboos University, Muscat. The official (and regularly updated) list of the birds of the Sultanate of Oman.

Sargeant, D. A., Eriksen, H. & Eriksen, J. 2008. *Birdwatching Guide to Oman (2nd edition)*. Al Roya Publishing, Muscat, Oman.

MIDDLE EAST
The Ornithological Society of the Middle East, OSME (www.osme.org) publishes *Sandgrouse* twice yearly. It contains topical articles, latest sightings and is usually where new birds for the region are published.

Blair, M., Preddy, S. & Al-Sirhan Alenezi, A. 2016. OSME Region List of Bird Taxa. v3.3-a, www.osme.org/orl

Christie, D. A., Shirihai, H. & Harris, A. 1996. *The Macmillan Birder's Guide to European and Middle Eastern Birds*. Macmillan, London.

Cramp, S., Simmons K. E. L. & Perrins, C. M. 1977–1994. *The Birds of the Western Palearctic* (BWP). Vols 1–9. Oxford University Press, Oxford.

Cramp, S., Simmons, K. E. L. & Perrins, C. M. 2004. *The Birds of the Western Palearctic: interactive* (BWPi). BirdGuides, Sheffield.

Delany, S., Scott, D., Dodman, T. & Stroud, D. 2009. (eds). *An Atlas of Wader Populations in Africa and Western Eurasia*. Wetlands International, Wageningen, The Netherlands.

Evans, M. I. 1994. *Important Bird Areas in the Middle East*. BirdLife Conservation Series No 2. BirdLife International, Cambridge.

Gill, F. & Donsker, D. (eds). 2016. IOC World Bird List (v 6.3). Available at www.worldbirdnames.org

Grimmett, R., Inskipp, C. & Inskipp, T. 1998. *Birds of the Indian Subcontinent*. Christopher Helm, London.

Jennings, M. C. 2010. Atlas of the Breeding Birds of Arabia. *Fauna of Arabia Vol. 25*.

Jonsson, L. 1992. *Birds of Europe with North Africa and the Middle East*. Christopher Helm, London.

Porter, R. F., Christensen, S. & Schiermacker-Hansen, P. 2006. [*Birds of the Middle East*] (in Arabic). Translated by Saeed Mohamed. SPNL. Beirut, Lebanon.

Rasmussen, P. C. & Anderton, J. C. 2012. *Birds of South Asia: The Ripley Guide*. Vols. 1 and 2. Smithsonian Institution, Michigan State University and Lynx Edicions, Washington DC, Michigan and Barcelona.

Redman, N., Stevenson, T. & Fanshawe, J. 2011. *Birds of the Horn of Africa*. 2nd Edition. Christopher Helm, London.

Svensson, L. 1992. *Identification Guide to European Passerines*. 4th Edition. Privately published.

Svensson, L., Mullarney, K. & Zetterström, D. 2009. *Collins Bird Guide*. 2nd Edition. HarperCollins, London.

van Duivendijk, N. 2010. *Advanced Bird ID Guide*. New Holland, London.

Williamson, K. 1976. *Identification for Ringers 1. The genera* Cettia, Locustella, Acrocephalus *and* Hippolais. Revised edition. BTO, Tring.

Williamson, K. 1976. *Identification for Ringers 2. The genus* Phylloscopus. Revised edition. BTO, Tring.

Williamson, K. 1976. *Identification for Ringers 3. The genus* Sylvia. Revised edition. BTO, Tring.

BAHRAIN
Hirschfeld, E. 1995. *Birds in Bahrain: A Study of their Migration Patterns 1990–92*. Hobby Publications, Dubai, UAE.

Mohamed, S. A. 1993. *Birds of Bahrain and the Arabian Gulf*. Bahrain Centre for Research Studies, Bahrain.

Nightingale, T. & Hill, M. 1993. *The Birds of Bahrain*. Immel, London.

CYPRUS
Flint, P. R. & Stewart, P. F. 1992. *The Birds of Cyprus*. 2nd Edition. BOU, Tring.

Richardson, C. 2009. *Birds of Cyprus Checklist 2003–2008 (2009)*. BirdLife Cyprus. (Earlier versions by other editors.) NB: Annual bird reports also published by BirdLife Cyprus.

IRAN
Mansoori, J. 2008. [*Field Guide to the Birds of Iran*] (in Farsi). Fahzanek Books, Tehran, Iran.

Khaleghizadeh, A., Roselaar, C., Scott, D. A. Tohidifar, M., Mlikovsky, J., Blair, M. & Kvartalnov, P. (in press). Birds of Iran: Annotated checklist of the bird species and subspecies. Iranian Research Institute of Plant Protection.

IRAQ

Salim, M. A., Porter, R. F., Schiermacker-Hansen, P., Christensen, S., & al-Jbour, S. 2006. [*Field Guide to the Birds of Iraq*.] (in Arabic). Nature Iraq and BirdLife International, Amman, Jordan.

Salim, M. A., Al-Sheikhly, O. F. Majeed, K. A. & Porter, R. F. 2012. An annotated checklist of the birds of Iraq. *Sandgrouse* 34: 4–43.

ISRAEL

Perlman, Y. & Meyrav, J. 2009. *Checklist of the Birds of Israel.* Soc. Prot. Nat. Israel/Israel Orn. Center, Tel-Aviv, Israel.

Sandgrouse 21(1). 1999. Special Levant Publication, including: Shirihai, H., Andrews, I. J., Kirwan, G. M. & Davidson, P. A checklist of the birds of Israel and Jordan.

Shirihai, H. 1996. *The Birds of Israel.* Academic Press, London.

JORDAN

Andrews, I. J. 1995. *The Birds of the Hashemite Kingdom of Jordan.* Privately published.

Andrews, I. J., Khoury, F. & Shirihai, H. 1999. Jordan Bird Report 1995–97. *Sandgrouse* 21(1): 10–35.

Sandgrouse 21(1). 1999. Special Levant Publication, including: Shirihai, H., Andrews, I. J., Kirwan, G. M. & Davidson, P. A checklist of the birds of Israel and Jordan.

KUWAIT

Gregory, G. 2005. *The Birds of the State of Kuwait.* Privately published.

Pope, M. & Zogaris, S., (eds). 2012. *Birds of Kuwait: A Comprehensive Visual Guide.* KUFPEC, Biodiversity East, Cyprus.

LEBANON

Ramadan-Jaradi, G., Bara, T. & Ramadan-Jaradi, M. 2008. Revised checklist of the birds of Lebanon 1999–2007. *Sandgrouse* 30(1): 22–69.

PALESTINE

al-Safadi, M. M. 2006. Observations on the breeding birds of the Gaza Strip, Palestine. *Sandgrouse* 28(1): 22–33.

QATAR

Eriksen, H. & J. and Gillespie, F. 2010. *Common Birds in Qatar.* Privately published.

Oldfield, C. & Oldfield, J. 1994. *A Birdwatcher's Guide to Qatar.* Privately published.

SAUDI ARABIA

Bundy, G., Connor, R. J. & Harrison, C. J. O. 1989. *Birds of the Eastern Province of Saudi Arabia.* H. F. & G. Witherby, London.

Jennings, M. C. 1981. *The Birds of Saudi Arabia: A Check-list.* Privately published.

SYRIA

Baumgart, W. 1995. *Die Vögel Syriens: Eine Übersicht.* Max Kasparek Verlag, Heidelberg, Germany.

Murdoch, D. A. & Betton, K. 2008. A checklist of the birds of Syria. OSME *Sandgrouse* Supplement 2. OSME, Sandy.

Syrian Society for Conservation of Wildlife & BirdLife International. 2009. [*Field Guide to the Birds of Syria*] (in Arabic). SSCW & BirdLife International, Damascus, Syria.

TURKEY

Green, I. & Moorhouse, N. 1995. *A Birdwatchers' Guide to Turkey.* Prion, Perry, UK.

Kirwan, G. M., Boyla, K. A., Castell, P., Demirci, B., Özen, M., Welch, H. & Marlow, T. 2008. *The Birds of Turkey.* Christopher Helm, London.

Porter, R. F., Christensen, S. & Schiermacker-Hansen, P. 2009 *Turkiyive Ortadogu'nun Kuslari* [Birds of Turkey and the Middle East]. Translated by Kerem Ali Boyla & Kazım Çapacı. Doğa Derneği, Turkey.

Roselaar, C. S. 1995. *Songbirds of Turkey: An Atlas of Biodiversity of Turkish Passerine Birds*. Pica Press, Robertsbridge, UK.

UNITED ARAB EMIRATES

Aspinall, S. J. 1996. *Status and Conservation of the Breeding Birds of the United Arab Emirates*. Hobby Publications, Warrington, UK and Dubai, UAE.

Aspinall, S. 2010. *The Breeding Birds of the United Arab Emirates (3rd edition)*. EAD, Abu Dhabi, UAE.

Aspinall, S., Javed, S. & Eriksen, H. & J. 2011. *Birds of the United Arab Emirates: a guide to common and important species*, EAD, Abu Dhabi, UAE.

Aspinall, S. & Porter, R. 2011. *Birds of the United Arab Emirates*. Christopher Helm, London.

Pedersen, T. & Aspinall, S. (compilers). 2010. EBRC Annotated checklist of the birds of the United Arab Emirates. *Sandgrouse* Supplement 3. OSME, Sandy, UK.

Richardson, C. 1990. *The Birds of the United Arab Emirates*. Hobby Publications, Warrington, UK and Dubai, UAE.

UAE Checklist. 2016: www.uaebirding.com (site contents continuously updated).

YEMEN

Multiple authors. 1987. [16 papers on the birds of northern Yemen]. *Sandgrouse* 9.

Multiple authors. 1996. [23 papers on the birds of southern Yemen and Socotra]. *Sandgrouse* 17.

SOCOTRA

Porter, R. F. & Suleiman, A. S. 2011. *Checklist of the Birds of the Socotra Archipelago*. www.osme.org.

Porter, R. F. & Suleiman, A. S. 2013. The populations and distribution of the breeding birds of the Socotra archipelago, Yemen: 1. Sandgrouse to Buntings. *Sandgrouse* 35(1): 43–81.

Porter, R. F. & Suleiman, A. S. 2014. The populations and distribution of the breeding birds of the Socotra archipelago, Yemen: 2. Shearwaters to Terns. *Sandgrouse* 36(1): 8–33.

Porter, R. F. & Suleiman, A. S. 2016. The Important Bird and Biodiversity Areas of the Socotra Archipelago, Yemen. *Sandgrouse* 38(2): 169–191.

CHECKLIST OF THE BIRDS OF OMAN

This checklist contains all the species recorded on the official Oman Bird List as at June 2017 (528 species). Non-naturalised escapes are excluded. It follows the order adopted by the International Ornithological Congress (IOC, see Introduction) but may differ from the sequence in the main text, as the latter has sometimes been changed for presentational reasons.

- ❏ Fulvous Whistling Duck *Dendrocygna bicolor*
- ❏ Lesser Whistling Duck *Dendrocygna javanica*
- ❏ Greylag Goose *Anser anser*
- ❏ Greater White-fronted Goose *Anser albifrons*
- ❏ Lesser White-fronted Goose *Anser erythropus*
- ❏ Red-breasted Goose *Branta ruficollis*
- ❏ Mute Swan *Cygnus olor*
- ❏ Bewick's Swan *Cygnus columbianus bewickii*
- ❏ Whooper Swan *Cygnus cygnus*
- ❏ Knob-billed Duck *Sarkidiornis melanotos*
- ❏ Egyptian Goose *Alopochen aegyptiaca*
- ❏ Common Shelduck *Tadorna tadorna*
- ❏ Ruddy Shelduck *Tadorna ferruginea*
- ❏ Cotton Pygmy Goose *Nettapus coromandelianus*
- ❏ Gadwall *Anas strepera*
- ❏ Eurasian Wigeon *Anas penelope*
- ❏ Mallard *Anas platyrhynchos*
- ❏ Northern Shoveler *Anas clypeata*
- ❏ Northern Pintail *Anas acuta*
- ❏ Garganey *Anas querquedula*
- ❏ Eurasian Teal *Anas crecca*
- ❏ Marbled Duck *Marmaronetta angustirostris*
- ❏ Red-crested Pochard *Netta rufina*
- ❏ Common Pochard *Aythya ferina*
- ❏ Ferruginous Duck *Aythya nyroca*
- ❏ Tufted Duck *Aythya fuligula*
- ❏ Red-breasted Merganser *Mergus serrator*
- ❏ Chukar Partridge *Alectoris chukar*
- ❏ Arabian Partridge *Alectoris melanocephala*
- ❏ Sand Partridge *Ammoperdix heyi*
- ❏ Grey Francolin *Francolinus pondicerianus*
- ❏ Common Quail *Coturnix coturnix*
- ❏ Harlequin Quail *Coturnix delegorguei*
- ❏ Wilson's Storm Petrel *Oceanites oceanicus*
- ❏ White-faced Storm Petrel *Pelagodroma marina*
- ❏ Black-bellied Storm Petrel *Fregetta tropica*
- ❏ Swinhoe's Storm Petrel *Oceanodroma monorhis*
- ❏ Matsudaira's Storm Petrel *Oceanodroma matsudairae*
- ❏ Streaked Shearwater *Calonectris leucomelas*
- ❏ Scopoli's Shearwater *Calonectris diomedea*
- ❏ Cory's Shearwater *Calonectris borealis*
- ❏ Wedge-tailed Shearwater *Puffinus pacificus*
- ❏ Sooty Shearwater *Puffinus griseus*
- ❏ Flesh-footed Shearwater *Puffinus carneipes*
- ❏ Persian Shearwater *Puffinus persicus*

- Jouanin's Petrel *Bulweria fallax*
- Little Grebe *Tachybaptus ruficollis*
- Great Crested Grebe *Podiceps cristatus*
- Black-necked Grebe *Podiceps nigricollis*
- Greater Flamingo *Phoenicopterus roseus*
- Lesser Flamingo *Phoeniconaias minor*
- Red-billed Tropicbird *Phaethon aethereus*
- Black Stork *Ciconia nigra*
- Abdim's Stork *Ciconia abdimii*
- White Stork *Ciconia ciconia*
- African Sacred Ibis *Threskiornis aethiopicus*
- Glossy Ibis *Plegadis falcinellus*
- Eurasian Spoonbill *Platalea leucorodia*
- African Spoonbill *Platalea alba*
- Eurasian Bittern *Botaurus stellaris*
- Little Bittern *Ixobrychus minutus*
- Yellow Bittern *Ixobrychus sinensis*
- Cinnamon Bittern *Ixobrychus cinnamomeus*
- Dwarf Bittern *Ixobrychus sturmii*
- Black-crowned Night Heron *Nycticorax nycticorax*
- Striated Heron *Butorides striata*
- Squacco Heron *Ardeola ralloides*
- Indian Pond Heron *Ardeola grayii*
- Chinese Pond Heron *Ardeola bacchus*
- Western Cattle Egret *Bubulcus ibis*
- Eastern Cattle Egret *Bubulcus coromandus*
- Grey Heron *Ardea cinerea*
- Black-headed Heron *Ardea melanocephala*
- Goliath Heron *Ardea goliath*
- Purple Heron *Ardea purpurea*
- Great Egret *Ardea alba*
- Intermediate Egret *Egretta intermedia*
- Black Heron *Egretta ardesiaca*
- Little Egret *Egretta garzetta*
- Western Reef Heron *Egretta gularis*
- Great White Pelican *Pelecanus onocrotalus*
- Pink-backed Pelican *Pelecanus rufescens*
- Dalmatian Pelican *Pelecanus crispus*
- Great Frigatebird *Fregata minor*
- Lesser Frigatebird *Fregata ariel*
- Cape Gannet *Morus capensis*
- Masked Booby *Sula dactylatra*
- Red-footed Booby *Sula sula*
- Brown Booby *Sula leucogaster*
- Great Cormorant *Phalacrocorax carbo*
- Socotra Cormorant *Phalacrocorax nigrogularis*
- Western Osprey *Pandion haliaetus*
- Black-winged Kite *Elanus caeruleus*
- Egyptian Vulture *Neophron percnopterus*
- European Honey Buzzard *Pernis apivorus*

- Crested Honey Buzzard *Pernis ptilorhynchus*
- Griffon Vulture *Gyps fulvus*
- Cinereous Vulture *Aegypius monachus*
- Lappet-faced Vulture *Torgos tracheliotos*
- Short-toed Snake Eagle *Circaetus gallicus*
- Lesser Spotted Eagle *Clanga pomarina*
- Greater Spotted Eagle *Clanga clanga*
- Booted Eagle *Hieraaetus pennatus*
- Tawny Eagle *Aquila rapax*
- Steppe Eagle *Aquila nipalensis*
- Eastern Imperial Eagle *Aquila heliaca*
- Golden Eagle *Aquila chrysaetos*
- Verreaux's Eagle *Aquila verreauxii*
- Bonelli's Eagle *Aquila fasciata*
- Shikra *Accipiter badius*
- Eurasian Sparrowhawk *Accipiter nisus*
- Northern Goshawk *Accipiter gentilis*
- Western Marsh Harrier *Circus aeruginosus*
- Hen Harrier *Circus cyaneus*
- Pallid Harrier *Circus macrourus*
- Montagu's Harrier *Circus pygargus*
- Black Kite *Milvus migrans*
- Yellow-billed Kite *Milvus aegyptius*
- Pallas's Fish Eagle *Haliaeetus leucoryphus*
- White-eyed Buzzard *Butastur teesa*
- Long-legged Buzzard *Buteo rufinus*
- Common Buzzard *Buteo buteo*
- Macqueen's Bustard *Chlamydotis macqueenii*
- Little Bustard *Tetrax tetrax*
- Water Rail *Rallus aquaticus*
- Corn Crake *Crex crex*
- White-breasted Waterhen *Amaurornis phoenicurus*
- Little Crake *Porzana parva*
- Baillon's Crake *Porzana pusilla*
- Spotted Crake *Porzana porzana*
- Ruddy-breasted Crake *Porzana fusca*
- Watercock *Gallicrex cinerea*
- Grey-headed Swamphen *Porphyrio poliocephalus*
- African Swamphen *Porphyrio madagascariensis*
- Allen's Gallinule *Porphyrio alleni*
- Common Moorhen *Gallinula chloropus*
- Lesser Moorhen *Gallinula angulata*
- Red-knobbed Coot *Fulica cristata*
- Eurasian Coot *Fulica atra*
- Demoiselle Crane *Grus virgo*
- Common Crane *Grus grus*
- Common Buttonquail *Turnix sylvaticus*
- Eurasian Stone-curlew *Burhinus oedicnemus*
- Spotted Thick-knee *Burhinus capensis*
- Great Stone-curlew *Esacus recurvirostris*

- ❏ Eurasian Oystercatcher *Haematopus ostralegus*
- ❏ Crab-plover *Dromas ardeola*
- ❏ Black-winged Stilt *Himantopus himantopus*
- ❏ Pied Avocet *Recurvirostra avosetta*
- ❏ Northern Lapwing *Vanellus vanellus*
- ❏ Spur-winged Lapwing *Vanellus spinosus*
- ❏ Grey-headed Lapwing *Vanellus cinereus*
- ❏ Red-wattled Lapwing *Vanellus indicus*
- ❏ Sociable Lapwing *Vanellus gregarius*
- ❏ White-tailed Lapwing *Vanellus leucurus*
- ❏ European Golden Plover *Pluvialis apricaria*
- ❏ Pacific Golden Plover *Pluvialis fulva*
- ❏ American Golden Plover *Pluvialis dominica*
- ❏ Grey Plover *Pluvialis squatarola*
- ❏ Common Ringed Plover *Charadrius hiaticula*
- ❏ Little Ringed Plover *Charadrius dubius*
- ❏ Kentish Plover *Charadrius alexandrinus*
- ❏ Lesser Sand Plover *Charadrius atrifrons*
- ❏ Greater Sand Plover *Charadrius leschenaultii*
- ❏ Caspian Plover *Charadrius asiaticus*
- ❏ Eurasian Dotterel *Charadrius morinellus*
- ❏ Greater Painted-snipe *Rostratula benghalensis*
- ❏ Pheasant-tailed Jacana *Hydrophasianus chirurgus*
- ❏ Eurasian Woodcock *Scolopax rusticola*
- ❏ Jack Snipe *Lymnocryptes minimus*
- ❏ Pin-tailed Snipe *Gallinago stenura*
- ❏ Great Snipe *Gallinago media*
- ❏ Common Snipe *Gallinago gallinago*
- ❏ Long-billed Dowitcher *Limnodromus scolopaceus*
- ❏ Asian Dowitcher *Limnodromus semipalmatus*
- ❏ Black-tailed Godwit *Limosa limosa*
- ❏ Bar-tailed Godwit *Limosa lapponica*
- ❏ Whimbrel *Numenius phaeopus*
- ❏ Slender-billed Curlew *Numenius tenuirostris*
- ❏ Eurasian Curlew *Numenius arquata*
- ❏ Far Eastern Curlew *Numenius madagascariensis*
- ❏ Spotted Redshank *Tringa erythropus*
- ❏ Common Redshank *Tringa totanus*
- ❏ Marsh Sandpiper *Tringa stagnatilis*
- ❏ Common Greenshank *Tringa nebularia*
- ❏ Lesser Yellowlegs *Tringa flavipes*
- ❏ Green Sandpiper *Tringa ochropus*
- ❏ Wood Sandpiper *Tringa glareola*
- ❏ Terek Sandpiper *Xenus cinereus*
- ❏ Common Sandpiper *Actitis hypoleucos*
- ❏ Ruddy Turnstone *Arenaria interpres*
- ❏ Great Knot *Calidris tenuirostris*
- ❏ Red Knot *Calidris canutus*
- ❏ Sanderling *Calidris alba*
- ❏ Little Stint *Calidris minuta*

- ❑ Temminck's Stint *Calidris temminckii*
- ❑ Long-toed Stint *Calidris subminuta*
- ❑ Baird's Sandpiper *Calidris bairdii*
- ❑ Pectoral Sandpiper *Calidris melanotos*
- ❑ Sharp-tailed Sandpiper *Calidris acuminata*
- ❑ Curlew Sandpiper *Calidris ferruginea*
- ❑ Dunlin *Calidris alpina*
- ❑ Broad-billed Sandpiper *Limicola falcinellus*
- ❑ Buff-breasted Sandpiper *Tryngites subruficollis*
- ❑ Ruff *Philomachus pugnax*
- ❑ Wilson's Phalarope *Steganopus tricolor*
- ❑ Red-necked Phalarope *Phalaropus lobatus*
- ❑ Red Phalarope *Phalaropus fulicarius*
- ❑ Cream-coloured Courser *Cursorius cursor*
- ❑ Collared Pratincole *Glareola pratincola*
- ❑ Black-winged Pratincole *Glareola nordmanni*
- ❑ Small Pratincole *Glareola lactea*
- ❑ Brown Noddy *Anous stolidus*
- ❑ Lesser Noddy *Anous tenuirostris*
- ❑ African Skimmer *Rynchops flavirostris*
- ❑ Indian Skimmer *Rynchops albicollis*
- ❑ Black-legged Kittiwake *Rissa tridactyla*
- ❑ Sabine's Gull *Xema sabini*
- ❑ Slender-billed Gull *Chroicocephalus genei*
- ❑ Brown-headed Gull *Chroicocephalus brunnicephalus*
- ❑ Black-headed Gull *Chroicocephalus ridibundus*
- ❑ Pallas's Gull *Ichthyaetus ichthyaetus*
- ❑ White-eyed Gull *Ichthyaetus leucophthalmus*
- ❑ Sooty Gull *Ichthyaetus hemprichii*
- ❑ Common Gull *Larus canus*
- ❑ Caspian Gull *Larus cachinnans*
- ❑ Lesser Black-backed Gull *Larus fuscus*
  - ❑ - Baltic Gull *Larus fuscus fuscus*
  - ❑ - Heuglin's Gull *Larus fuscus heuglini*
  - ❑ - Steppe Gull *Larus fuscus barabensis*
- ❑ Gull-billed Tern *Gelochelidon nilotica*
- ❑ Caspian Tern *Hydroprogne caspia*
- ❑ Greater Crested Tern *Sterna bergii*
- ❑ Lesser Crested Tern *Sterna bengalensis*
- ❑ Sandwich Tern *Sterna sandvicensis*
- ❑ Little Tern *Sternula albifrons*
- ❑ Saunders's Tern *Sternula saundersi*
- ❑ Bridled Tern *Onychoprion anaethetus*
- ❑ Sooty Tern *Onychoprion fuscatus*
- ❑ Roseate Tern *Sterna dougallii*
- ❑ Common Tern *Sterna hirundo*
- ❑ White-cheeked Tern *Sterna repressa*
- ❑ Arctic Tern *Sterna paradisaea*
- ❑ Whiskered Tern *Chlidonias hybrida*
- ❑ White-winged Tern *Chlidonias leucopterus*

- Black Tern *Chlidonias niger*
- South Polar Skua *Stercorarius maccormicki*
- Brown Skua *Stercorarius antarcticus*
- Pomarine Skua *Stercorarius pomarinus*
- Arctic Skua *Stercorarius parasiticus*
- Long-tailed Skua *Stercorarius longicaudus*
- Pin-tailed Sandgrouse *Pterocles alchata*
- Chestnut-bellied Sandgrouse *Pterocles exustus*
- Spotted Sandgrouse *Pterocles senegallus*
- Crowned Sandgrouse *Pterocles coronatus*
- Lichtenstein's Sandgrouse *Pterocles lichtensteinii*
- Rock Dove *Columba livia*
- Stock Dove *Columba oenas*
- Common Wood Pigeon *Columba palumbus*
- European Turtle Dove *Streptopelia turtur*
- Oriental Turtle Dove *Streptopelia orientalis*
- Eurasian Collared Dove *Streptopelia decaocto*
- African Collared Dove *Streptopelia roseogrisea*
- Red Turtle Dove *Streptopelia tranquebarica*
- Laughing Dove *Spilopelia senegalensis*
- Namaqua Dove *Oena capensis*
- Bruce's Green Pigeon *Treron waalia*
- Great Spotted Cuckoo *Clamator glandarius*
- Jacobin Cuckoo *Clamator jacobinus*
- Asian Koel *Eudynamys scolopaceus*
- Diederik Cuckoo *Chrysococcyx caprius*
- Grey-bellied Cuckoo *Cacomantis passerinus*
- Common Hawk-Cuckoo *Hierococcyx varius*
- Common Cuckoo *Cuculus canorus*
- Western Barn Owl *Tyto alba*
- Pallid Scops Owl *Otus brucei*
- Arabian Scops Owl *Otus pamelae*
- Eurasian Scops Owl *Otus scops*
- Pharaoh Eagle-Owl *Bubo ascalaphus*
- Spotted Eagle-Owl *Bubo africanus*
- Desert Owl *Strix hadorami*
- Omani Owl *Strix butleri*
- Little Owl *Athene noctua*
- Long-eared Owl *Asio otus*
- Short-eared Owl *Asio flammeus*
- European Nightjar *Caprimulgus europaeus*
- Egyptian Nightjar *Caprimulgus aegyptius*
- Nubian Nightjar *Caprimulgus nubicus*
- Alpine Swift *Tachymarptis melba*
- Common Swift *Apus apus*
- Pallid Swift *Apus pallidus*
- Forbes-Watson's Swift *Apus berliozi*
- Pacific Swift *Apus pacificus*
- Little Swift *Apus affinis*
- Indian Roller *Coracias benghalensis*

- ❏ Lilac-breasted Roller *Coracias caudatus*
- ❏ European Roller *Coracias garrulus*
- ❏ White-throated Kingfisher *Halcyon smyrnensis*
- ❏ Grey-headed Kingfisher *Halcyon leucocephala*
- ❏ Collared Kingfisher *Todiramphus chloris*
- ❏ Malachite Kingfisher *Corythornis cristata*
- ❏ Common Kingfisher *Alcedo atthis*
- ❏ Pied Kingfisher *Ceryle rudis*
- ❏ White-throated Bee-eater *Merops albicollis*
- ❏ Green Bee-eater *Merops orientalis*
- ❏ Blue-cheeked Bee-eater *Merops persicus*
- ❏ European Bee-eater *Merops apiaster*
- ❏ Eurasian Hoopoe *Upupa epops*
- ❏ Eurasian Wryneck *Jynx torquilla*
- ❏ Lesser Kestrel *Falco naumanni*
- ❏ Common Kestrel *Falco tinnunculus*
- ❏ Amur Falcon *Falco amurensis*
- ❏ Eleonora's Falcon *Falco eleonorae*
- ❏ Sooty Falcon *Falco concolor*
- ❏ Merlin *Falco columbarius*
- ❏ Eurasian Hobby *Falco subbuteo*
- ❏ Lanner Falcon *Falco biarmicus*
- ❏ Saker Falcon *Falco cherrug*
- ❏ Peregrine Falcon *Falco peregrinus*
- ❏ Barbary Falcon *Falco pelegrinoides*
- ❏ Alexandrine Parakeet *Psittacula eupatria*
- ❏ Rose-ringed Parakeet *Psittacula krameri*
- ❏ Black-crowned Tchagra *Tchagra senegala*
- ❏ Brown Shrike *Lanius cristatus*
- ❏ Red-backed Shrike *Lanius collurio*
- ❏ Isabelline Shrike *Lanius isabellinus*
- ❏ Red-tailed Shrike *Lanius phoenicuroides*
- ❏ Bay-backed Shrike *Lanius vittatus*
- ❏ Long-tailed Shrike *Lanius schach*
- ❏ Lesser Grey Shrike *Lanius minor*
- ❏ Southern Grey Shrike *Lanius meridionalis*
- ❏ Steppe Grey Shrike *Lanius pallidirostris*
- ❏ Woodchat Shrike *Lanius senator*
- ❏ Masked Shrike *Lanius nubicus*
- ❏ Eurasian Golden Oriole *Oriolus oriolus*
- ❏ Black-naped Oriole *Oriolus chinensis*
- ❏ Black Drongo *Dicrurus macrocercus*
- ❏ Ashy Drongo *Dicrurus leucophaeus*
- ❏ African Paradise Flycatcher *Terpsiphone viridis*
- ❏ Eurasian Magpie *Pica pica*
- ❏ House Crow *Corvus splendens*
- ❏ Brown-necked Raven *Corvus ruficollis*
- ❏ Fan-tailed Raven *Corvus rhipidurus*
- ❏ Grey Hypocolius *Hypocolius ampelinus*
- ❏ Black-headed Penduline Tit *Remiz macronyx*

- Greater Hoopoe-Lark *Alaemon alaudipes*
- Thick-billed Lark *Ramphocoris clotbey*
- Desert Lark *Ammomanes deserti*
- Bar-tailed Lark *Ammomanes cinctura*
- Black-crowned Sparrow-Lark *Eremopterix nigriceps*
- Singing Bush Lark *Mirafra cantillans*
- Oriental Skylark *Alauda gulgula*
- Eurasian Skylark *Alauda arvensis*
- Crested Lark *Galerida cristata*
- Blanford's Lark *Calandrella blanfordi*
- Greater Short-toed Lark *Calandrella brachydactyla*
- Bimaculated Lark *Melanocorypha bimaculata*
- Calandra Lark *Melanocorypha calandra*
- Dunn's Lark *Eremalauda dunni*
- Lesser Short-toed Lark *Alaudala rufescens*
- White-eared Bulbul *Pycnonotus leucotis*
- Red-vented Bulbul *Pycnonotus cafer*
- White-spectacled Bulbul *Pycnonotus xanthopygos*
- Grey-throated Martin *Riparia chinensis*
- Sand Martin *Riparia riparia*
- Pale Martin *Riparia diluta*
- Barn Swallow *Hirundo rustica*
- Wire-tailed Swallow *Hirundo smithii*
- Eurasian Crag Martin *Ptyonoprogne rupestris*
- Pale Crag Martin *Ptyonoprogne obsoleta*
- Common House Martin *Delichon urbicum*
- Lesser Striped Swallow *Cecropis abyssinica*
- Red-rumped Swallow *Cecropis daurica*
- Streak-throated Swallow *Petrochelidon fluvicola*
- Cetti's Warbler *Cettia cetti*
- Streaked Scrub Warbler *Scotocerca inquieta*
- Willow Warbler *Phylloscopus trochilus*
- Common Chiffchaff *Phylloscopus collybita*
  - - Scandinavian Chiffchaff *Phylloscopus collybita abietinus*
  - - Siberian Chiffchaff *Phylloscopus collybita tristis*
- Plain Leaf Warbler *Phylloscopus neglectus*
- Eastern Bonelli's Warbler *Phylloscopus orientalis*
- Wood Warbler *Phylloscopus sibilatrix*
- Dusky Warbler *Phylloscopus fuscatus*
- Yellow-browed Warbler *Phylloscopus inornatus*
- Hume's Leaf Warbler *Phylloscopus humei*
- Arctic Warbler *Phylloscopus borealis*
- Green Warbler *Phylloscopus nitidus*
- Greenish Warbler *Phylloscopus trochiloides*
- Great Reed Warbler *Acrocephalus arundinaceus*
- Clamorous Reed Warbler *Acrocephalus stentoreus*
- Moustached Warbler *Acrocephalus melanopogon*
- Sedge Warbler *Acrocephalus schoenobaenus*
- Paddyfield Warbler *Acrocephalus agricola*
- Blyth's Reed Warbler *Acrocephalus dumetorum*

- Eurasian Reed Warbler *Acrocephalus scirpaceus*
- Marsh Warbler *Acrocephalus palustris*
- Thick-billed Warbler *Iduna aedon*
- Booted Warbler *Iduna caligata*
- Sykes's Warbler *Iduna rama*
- Eastern Olivaceous Warbler *Iduna pallida*
- Upcher's Warbler *Hippolais languida*
- Olive-tree Warbler *Hippolais olivetorum*
- Icterine Warbler *Hippolais icterina*
- Common Grasshopper Warbler *Locustella naevia*
- River Warbler *Locustella fluviatilis*
- Savi's Warbler *Locustella luscinioides*
- Zitting Cisticola *Cisticola juncidis*
- Graceful Prinia *Prinia gracilis*
- Arabian Babbler *Turdoides squamiceps*
- Eurasian Blackcap *Sylvia atricapilla*
- Garden Warbler *Sylvia borin*
- Barred Warbler *Sylvia nisoria*
- Lesser Whitethroat *Sylvia curruca*
- Hume's Whitethroat *Sylvia althaea*
- Eastern Orphean Warbler *Sylvia crassirostris*
- Arabian Warbler *Sylvia leucomelaena*
- Asian Desert Warbler *Sylvia nana*
- Common Whitethroat *Sylvia communis*
- Sardinian Warbler *Sylvia melanocephala*
- Ménétriés's Warbler *Sylvia mystacea*
- Oriental White-eye *Zosterops palpebrosus*
- Abyssinian White-eye *Zosterops abyssinicus*
- Bank Myna *Acridotheres ginginianus*
- Common Myna *Acridotheres tristis*
- Chestnut-tailed Starling *Sturnia malabarica*
- Brahminy Starling *Sturnia pagodarum*
- Rosy Starling *Pastor roseus*
- Common Starling *Sturnus vulgaris*
- Wattled Starling *Creatophora cinerea*
- Violet-backed Starling *Cinnyricinclus leucogaster*
- Tristram's Starling *Onychognathus tristramii*
- White's Thrush *Zoothera aurea*
- Ring Ouzel *Turdus torquatus*
- Eyebrowed Thrush *Turdus obscurus*
- Black-throated Thrush *Turdus atrogularis*
- Red-throated Thrush *Turdus ruficollis*
- Dusky Thrush *Turdus eunomus*
- Song Thrush *Turdus philomelos*
- Mistle Thrush *Turdus viscivorus*
- Rufous-tailed Scrub Robin *Cercotrichas galactotes*
- Black Scrub Robin *Cercotrichas podobe*
- Spotted Flycatcher *Muscicapa striata*
- Asian Brown Flycatcher *Muscicapa dauurica*
- Blue-and-white Flycatcher *Cyanoptila cyanomelana*

- European Robin *Erithacus rubecula*
- Bluethroat *Luscinia svecica*
- Thrush Nightingale *Luscinia luscinia*
- Common Nightingale *Luscinia megarhynchos*
- White-throated Robin *Irania gutturalis*
- European Pied Flycatcher *Ficedula hypoleuca*
- Semicollared Flycatcher *Ficedula semitorquata*
- Red-breasted Flycatcher *Ficedula parva*
- Taiga Flycatcher *Ficedula albicilla*
- Eversmann's Redstart *Phoenicurus erythronotus*
- Black Redstart *Phoenicurus ochruros*
- Common Redstart *Phoenicurus phoenicurus*
- Common Rock Thrush *Monticola saxatilis*
- Blue Rock Thrush *Monticola solitarius*
- Whinchat *Saxicola rubetra*
- European Stonechat *Saxicola rubicola*
- Siberian Stonechat *Saxicola maurus*
- Pied Bush Chat *Saxicola caprata*
- Northern Wheatear *Oenanthe oenanthe*
- Isabelline Wheatear *Oenanthe isabellina*
- Hooded Wheatear *Oenanthe monacha*
- Desert Wheatear *Oenanthe deserti*
- Black-eared Wheatear *Oenanthe hispanica*
- Cyprus Wheatear *Oenanthe cypriaca*
- Pied Wheatear *Oenanthe pleschanka*
- Blackstart *Oenanthe melanura*
- Variable Wheatear *Oenanthe picata*
- White-crowned Wheatear *Oenanthe leucopyga*
- Hume's Wheatear *Oenanthe albonigra*
- Finsch's Wheatear *Oenanthe finschii*
- Mourning Wheatear *Oenanthe lugens*
- Arabian Wheatear *Oenanthe lugentoides*
- Kurdish Wheatear *Oenanthe xanthoprymna*
- Red-tailed Wheatear *Oenanthe chrysopygia*
- Nile Valley Sunbird *Hedydipna metallica*
- Palestine Sunbird *Cinnyris osea*
- Shining Sunbird *Cinnyris habessinicus*
- Purple Sunbird *Cinnyris asiaticus*
- House Sparrow *Passer domesticus*
- Spanish Sparrow *Passer hispaniolensis*
- Eurasian Tree Sparrow *Passer montanus*
- Pale Rockfinch *Carpospiza brachydactyla*
- Yellow-throated Sparrow *Gymnoris xanthocollis*
- Rüppell's Weaver *Ploceus galbula*
- African Silverbill *Euodice cantans*
- Indian Silverbill *Lonchura malabarica*
- Scaly-breasted Munia *Lonchura punctulata*
- Radde's Accentor *Prunella ocularis*
- Black-throated Accentor *Prunella atrogularis*
- Forest Wagtail *Dendronanthus indicus*

- ❏ Western Yellow Wagtail *Motacilla flava*
  - ❏ - Yellow-headed Wagtail *Motacilla flava lutea*
  - ❏ - Blue-headed Wagtail *Motacilla flava flava*
  - ❏ - Sykes's Wagtail *Motacilla flava beema*
  - ❏ - Ashy-headed Wagtail *Motacilla flava cinereocapilla*
  - ❏ - White-headed Wagtail *Motacilla flava leucocephala*
  - ❏ - Black-headed Wagtail *Motacilla flava feldegg*
  - ❏ - Grey-headed Wagtail *Motacilla flava thunbergi*
- ❏ Eastern Yellow Wagtail *Motacilla tschutschensis*
- ❏ Citrine Wagtail *Motacilla citreola*
- ❏ Grey Wagtail *Motacilla cinerea*
- ❏ White Wagtail *Motacilla alba*
  - ❏ - Masked Wagtail *Motacilla alba personata*
  - ❏ - Amur Wagtail *Motacilla alba leucopsis*
- ❏ Golden Pipit *Tmetothylacus tenellus*
- ❏ Richard's Pipit *Anthus richardi*
- ❏ Blyth's Pipit *Anthus godlewskii*
- ❏ Tawny Pipit *Anthus campestris*
- ❏ Long-billed Pipit *Anthus similis*
- ❏ Meadow Pipit *Anthus pratensis*
- ❏ Tree Pipit *Anthus trivialis*
- ❏ Olive-backed Pipit *Anthus hodgsoni*
- ❏ Red-throated Pipit *Anthus cervinus*
- ❏ Buff-bellied Pipit *Anthus rubescens*
- ❏ Water Pipit *Anthus spinoletta*
- ❏ Common Chaffinch *Fringilla coelebs*
- ❏ Brambling *Fringilla montifringilla*
- ❏ Trumpeter Finch *Bucanetes githagineus*
- ❏ Common Rosefinch *Carpodacus erythrinus*
- ❏ Arabian Golden-winged Grosbeak *Rhynchostruthus percivali*
- ❏ Yemen Serin *Crithagra menachensis*
- ❏ European Goldfinch *Carduelis carduelis*
- ❏ Eurasian Siskin *Carduelis spinus*
- ❏ Corn Bunting *Emberiza calandra*
- ❏ Pine Bunting *Emberiza leucocephalos*
- ❏ Grey-necked Bunting *Emberiza buchanani*
- ❏ Cinereous Bunting *Emberiza cineracea*
- ❏ Ortolan Bunting *Emberiza hortulana*
- ❏ Cretzschmar's Bunting *Emberiza caesia*
- ❏ Striolated Bunting *Emberiza striolata*
- ❏ Cinnamon-breasted Bunting *Emberiza tahapisi*
- ❏ Little Bunting *Emberiza pusilla*
- ❏ Rustic Bunting *Emberiza rustica*
- ❏ Yellow-breasted Bunting *Emberiza aureola*
- ❏ Black-headed Bunting *Emberiza melanocephala*
- ❏ Red-headed Bunting *Emberiza bruniceps*
- ❏ Common Reed Bunting *Emberiza schoeniclus*

INDEX

A

Accentor, Black-throated 230
  Radde's 230
Accipiter badius 64
  gentilis 64
  nisus 64
Acridotheres ginginianus 200
  tristis 200
Acrocephalus agricola 186
  arundinaceus 186
  dumetorum 188
  melanopogon 186
  palustris 188
  schoenobaenus 186
  scirpaceus 188
  scirpaceus fuscus 188
  stentoreus 186
  stentoreus brunnescens 186
Actitis hypoleucos 102
Aegypius monachus 52
Alaemon alaudipes 170
Alauda arvensis 172
  gulgula 172
Alaudala rufescens 174
Alcedo atthis 156
Alectoris chukar 26
  melanocephala 26
Alopochen aegyptiaca 18
Amandava amandava 228
Amaurornis phoenicurus 80
Ammomanes cinctura 170
  deserti 170
Ammoperdix heyi 26
Anas acuta 22
  clypeata 22
  crecca 22
  penelope 20
  platyrhynchos 22
  poecilorhyncha 20
  querquedula 22
  strepera 20
Anous stolidus 114
  tenuirostris 114
Anser albifrons 18
  anser 18
  anser rubrirostris 18
  erythropus 18
Anthus campestris 236
  cervinus 238
  godlewskii 236
  hodgsoni 238
  pratensis 238
  richardi 236
  rubescens 238
  similis 236
  spinoletta 238
  trivialis 238

Apus affinis 152
  apus 152
  apus pekinensis 152
  berliozi 152
  pacificus 152
  pallidus 152
Aquila chrysaetos 60
  fasciata 62
  heliaca 60
  nipalensis 60
  rapax 58
  rapax vindhiana 58
  verreauxii 62
Ardea alba 46
  cinerea 44
  goliath 44
  melanocephala 44
  purpurea 44
Ardenna carneipes 32
  grisea 32
  pacifica 32
Ardeola bacchus 42
  grayii 42
  ralloides 42
Arenaria interpres 102
Asio flammeus 150
  otus 150
Athene noctua 150
  noctua lilith 150
Avadavat, Red 228
Avocet, Pied 84
Aythya ferina 24
  fuligula 24
  nyroca 24

B

Babbler, Arabian 164
Bee-eater, Blue-cheeked 158
  European 158
  Green 158
  Little Green 158
  White-throated 158
Bishop, Southern Red 226
  Yellow-crowned 226
Bittern, Cinnamon 40
  Dwarf 40
  Eurasian 40
  Little 40
  Yellow 40
Blackcap, Eurasian 192
Blackstart 214
Bluethroat 206
Booby, Brown 50
  Masked 50
  Red-footed 50
Botaurus stellaris 40
Brambling 240

Branta ruficollis 18
Bubo *africanus* 148
   africanus milesi 148
   ascalaphus 148
Bubulcus coromandus 42
   ibis 42
Bucanetes githagineus 240
Bulbul, Red-vented 166
   White-cheeked 166
   White-eared 166
   White-spectacled 166
   Yellow-vented 166
Bulweria fallax 32
Bunting, African Rock 242
   Black-headed 246
   Cinereous 244
   Cinnamon-breasted 242
   Common Reed 246
   Corn 242
   Cretzschmar's 244
   Eastern Cinereous 242
   Grey-necked 244
   Little 244
   Ortolan 244
   Pine 242
   Red-headed 246
   Rustic 246
   Striolated 242
   Yellow-breasted 246
Burhinus capensis 76
   oedicnemus 76
Bustard, Houbara 76
   Little 76
   Macqueen's 76
Butastur teesa 68
Buteo buteo 68
   buteo vulpinus 68
   rufinus 68
Butorides striata 42
Buttonquail, Common 24
   Kurrichane 24
Buzzard, Common 68
   Crested Honey 54
   European Honey 54
   Long-legged 68
   Steppe 68
   White-eyed 68

C

Cacomantis merulinus 144
   passerinus 144
Calandrella blanfordi 174
   brachydactyla 174
   eremica 174
Calidris acuminata 108
   alba 104
   alpina 106
   bairdii 108
   canutus 106
   falcinellus 106

   ferruginea 106
   melanotos 108
   minuta 104
   pugnax 110
   subminuta 104
   subruficollis 108
   temminckii 104
   tenuirostris 106
Calonectris borealis 30
   diomedea 30
   leucomelas 30
Caprimulgus aegyptius 144
   europaeus 144
   nubicus 144
Carduelis carduelis 240
Carpodacus erythrinus 240
Carpospiza brachydactyla 224
Cecropis abyssinica 178
   daurica 178
Cercotrichas galactotes 206
   podobe 206
Ceryle rudis 156
Cettia cetti 180
Charadrius alexandrinus 90
   asiaticus 92
   dubius 90
   hiaticula 90
   leschenaultii 92
   mongolus 92
   morinellus 90
Chat, Pied Bush 210
Chiffchaff, Common 182
   Scandinavian 182
   Siberian 182
Chlamydotis macqueenii 76
   undulata 76
Chlidonias hybrida 130
   leucopterus 130
   niger 130
Chroicocephalus brunnicephalus 116
   genei 116
   ridibundus 116
Chrysococcyx caprius 142
Ciconia abdimii 36
   ciconia 36
   nigra 36
Cinnyricinclus leucogaster 198
Cinnyris asiaticus 222
   habessinicus 222
   osea 222
Circaetus gallicus 62
Circus aeruginosus 66
   cyaneus 66
   macrourus 66
   pygargus 66
Cisticola juncidis 180
Cisticola, Zitting 180
Clamator glandarius 142
   jacobinus 142
Clanga clanga 58

pomarina 58
Columba livia 136
   oenas 136
   palumbus 136
Coot, Crested 82
   Eurasian 82
   Red-knobbed 82
Coracias benghalensis 154
   caudatus 154
   caudatus lorti 154
   garrulus 154
Cormorant, Great 50
   Socotra 50
Corvus rhipidurus 168
   ruficollis 168
   splendens 168
Corythornis cristata 156
Coturnix coturnix 26
   delegorguei 26
Courser, Cream-coloured 112
Crab-plover 84
Crake, Baillon's 78
   Corn 80
   Little 78
   Ruddy-breasted 78
   Spotted 78
Crane, Common 36
   Demoiselle 36
Creatophora cinerea 198
Crex crex 80
Crithagra menachensis 240
Crow, House 168
   Indian House 168
Cuckoo, Common 142
   Didric 142
   Diederik 142
   Great Spotted 142
   Grey-bellied 144
   Jacobin 142
   Pied 142
   Plaintive 144
Cuculus canorus 142
Curlew, Eastern 98
   Eurasian 98
   Far Eastern 98
   Slender-billed 98
Cursorius cursor 112
Cyanoptila cyanomelana 222
Cygnus columbianus bewickii 16
   cygnus 16
   olor 16

D

Delichon urbicum 176
Dendrocygna bicolor 16
   javanica 16
Dendronanthus indicus 230
Dicrurus leucophaeus 164
   macrocercus 164
Dikkop, Spotted 76

Dotterel, Eurasian 90
Dove, African Collared 138
   Barbary 138
   Eurasian Collared 138
   European Turtle 136
   Laughing 138
   Namaqua 138
   Oriental Turtle 136
   Palm 138
   Red Turtle 138
   Rock 136
   Rufous Turtle 136
   Stock 136
Dowitcher, Asian 96
   Long-billed 96
Dromas ardeola 84
Drongo, Ashy 164
   Black 164
Duck, Comb 16
   Ferruginous 24
   Fulvous Whistling 16
   Indian Spot-billed 20
   Knob-billed 16
   Lesser Whistling 16
   Marbled 20
   Tufted 24
Dunlin 106

E

Eagle, Bonelli's 62
   Booted 62
   Eastern Imperial 60
   Golden 60
   Greater Spotted 58
   Indian Tawny 58
   Lesser Spotted 58
   Pallas's Fish 64
   Short-toed 62
   Short-toed Snake 62
   Steppe 60
   Tawny 58
   Verreaux's 62
Eagle-Owl, Arabian Spotted 148
   Desert 148
   Pharaoh 148
   Spotted 148
Egret, Eastern Cattle 42
   Great 46
   Great White 46
   Intermediate 46
   Little 46
   Western Cattle 42
   Western Great 46
   Western Reef 46
Egretta ardesiaca 46
   garzetta 46
   gularis 46
   gularis schistacea 46
   intermedia 46
Elanus caeruleus 56

Emberiza aureola 246
  bruniceps 246
  buchanani 244
  caesia 244
  calandra 242
  cineracea 244
  cineracea semenowi 242
  hortulana 244
  leucocephala 242
  melanocephala 246
  pusilla 244
  rustica 246
  schoeniclus 246
  striolata 242
  tahapisi 242
Eremalauda dunni 170
  eremodites 170
Eremopterix nigriceps 172
Erithacus rubecula 206
Esacus recurvirostris 76
Estrilda astrild 228
Eudynamys scolopaceus 142
Euodice cantans 228
  malabarica 228
Euplectes afer 226
  orix 226

F

Falco amurensis 72
  biarmicus 74
  cherrug 74
  columbarius 70
  concolor 72
  eleonorae 72
  naumanni 70
  pelegrinoides 74
  peregrinus 74
  subbuteo 72
  tinnunculus 70
Falcon, Amur 72
  Barbary 74
  Eleonora's 72
  Lanner 74
  Peregrine 74
  Saker 74
  Sooty 72
Ficedula albicilla 220
  hypoleuca 220
  parva 220
  semitorquata 220
Finch, Trumpeter 240
Flamingo, Greater 34
  Lesser 34
Flycatcher, African Paradise 166
  Asian Brown 220
  Blue-and-white 222
  European Pied 220
  Red-breasted 220
  Semicollared 220
  Spotted 220

  Taiga 220
Francolin, Grey 26
Francolinus pondicerianus 26
Fregata ariel 48
  minor 48
Fregetta tropica 28
Frigatebird, Great 48
  Lesser 48
Fringilla montifringilla 240
Fulica atra 82
  cristata 82

G

Gadwall 20
Galerida cristata 172
Gallicrex cinerea 80
Gallinago gallinago 94
  media 94
  stenura 94
Gallinula angulata 82
  chloropus 82
Gallinule, Allen's 82
Gannet, Cape 50
Garganey 22
Gelochelidon nilotica 124
Glareola lactea 112
  maldivarum 112
  nordmanni 112
  pratincola 112
Godwit, Bar-tailed 96
  Black-tailed 96
Goldfinch, European 240
Goose, Cotton Pygmy 18
  Eastern Greylag 18
  Egyptian 18
  Greater White-fronted 18
  Greylag 18
  Lesser White-fronted 18
  Red-breasted 18
Goshawk, Northern 64
Grackle, Tristram's 200
Grebe, Black-necked 34
  Great Crested 34
  Little 34
Greenshank, Common 100
Grosbeak, Arabian Golden-winged 242
Grus grus 36
  virgo 36
Guineafowl, Helmeted 24
Gull, Baltic 120
  Black-headed 116
  Brown-headed 116
  Caspian 120, 122
  Common 118
  Great Black-headed 118
  Heuglin's 120, 122
  Lesser Black-backed 120
  Pallas's 118
  Sabine's 116
  Siberian 120

Slender-billed 116
Sooty 118
Steppe 120, 122
White-eyed 118
Gymnoris xanthocollis 224
Gyps fulvus 52

H

Haematopus ostralegus 84
Halcyon leucocephala 156
    smyrnensis 156
Haliaeetus leucoryphus 64
Haliastur indus 56
Harrier, Hen 66
    Montagu's 66
    Pallid 66
    Western Marsh 66
Hawk-Cuckoo, Common 144
    Indian 144
Hedydipna metallica 222
Hemipode, Andalusian 24
Heron, Black 46
    Black-crowned Night 44
    Black-headed 44
    Chinese Pond 42
    Goliath 44
    Grey 44
    Indian Pond 42
    Indian Reef 46
    Purple 44
    Squacco 42
    Striated 42
    Western Reef 46
Hieraaetus pennatus 62
Hierococcyx varius 144
Himantopus himantopus 84
Hippolais icterina 192
    languida 190
    olivetorum 190
Hirundo rustica 178
    smithii 178
Hobby, Eurasian 72
Hoopoe, Eurasian 154
Hoopoe-Lark, Greater 170
Hydrophasianus chirurgus 88
Hydroprogne caspia 124
Hypocolius ampelinus 166
Hypocolius, Grey 166
Ibis, African Sacred 38
    Glossy 38

I

Ichthyaetus hemprichii 118
    ichthyaetus 118
    leucophthalmus 118
Iduna aedon 190
    caligata 190
    pallida 190
    rama 190
Irania gutturalis 206

Ixobrychus cinnamomeus 40
    minutus 40
    sinensis 40
    sturmii 40
Jacana, Pheasant-tailed 88
Jaeger, Long-tailed 132
    Parasitic 132
    Pomarine 132
Jynx torquilla 154

K

Kestrel, Common 70
    Lesser 70
Kingfisher, Collared 156
    Common 156
    Grey-headed 156
    Malachite 156
    Mangrove 156
    Pied 156
    White-throated 156
Kite, Black 56
    Black-eared 56
    Black-winged 56
    Brahminy 56
    Yellow-billed 56
Kittiwake, Black-legged 116
Knot, Great 106
    Red 106
Koel, Asian 142

L

Lamprotornis superbus 198
Lanius collurio 160
    cristata 160
    isabellinus 160
    meridionalis 162
    minor 162
    nubicus 160
    pallidirostris 162
    phoenicuroides 160
    schach 162
    senator 160
    vittatus 162
Lapwing, Grey-headed 86
    Northern 86
    Red-wattled 86
    Sociable 86
    Spur-winged 86
    White-tailed 86
Lark, Arabian 170
    Bar-tailed 170
    Bimaculated 174
    Blanford's 174
    Calandra 174
    Crested 172
    Desert 170
    Dunn's 170
    Greater Short-toed 174
    Lesser Short-toed 174
    Red-capped 174

Rufous-capped 174
Singing Bush 172
Thick-billed 170
Larus cachinnans 120
   canus 118
   fuscus 120
   fuscus barabensis 120
   fuscus fuscus 120
   fuscus heuglini 120
Limnodromus scolopaceus 96
   semipalmatus 96
Limosa lapponica 96
   limosa 96
Locustella fluviatilis 188
   luscinioides 188
   naevia 188
Lonchura atricapilla 228
   malacca 228
   punctulata 228
Luscinia luscinia 208
   megarhynchos 208
   megarhynchos golzii 208
   svecica 206
Lymnocryptes minimus 94

M

Magpie, Eurasian 168
Mallard 22
Marmaronetta angustirostris 20
Martin, African Rock 176
   Brown-throated 176
   Common House 176
   Eurasian Crag 176
   Grey-throated 176
   Pale 176
   Pale Crag 176
   Sand 176
Melanocorypha bimaculata 174
   calandra 174
Merganser, Red-breasted 22
Mergus serrator 22
Merlin 70
Merops albicollis 158
   apiaster 158
   orientalis 158
   persicus 158
Milvus aegyptius 56
   migrans 56
   migrans lineatus 56
Mirafra cantillans 172
Monticola saxatilis 204
   solitarius 204
Moorhen, Common 82
   Lesser 82
Morus capensis 50
Motacilla alba 232
   alba leucopsis 232
   alba personata 232
   cinerea 230
   citreola 230

   citreola calcarata 230
   flava 234
   flava beema 234
   flava feldegg 234
   flava leucocephala 234
   flava lutea 234
   flava thunbergi 234
   tschutschensis 232
   tschutschensis taivana 232
Munia, Chestnut 228
   Scaly-breasted 228
   Tricoloured 228
Muscicapa dauurica 220
   striata 220
Mycteria ibis 36
Myna, Bank 200
   Common 200

N

Neophron percnopterus 52
Netta rufina 24
Nettapus coromandelianus 18
Nightingale, Common 208
   Eastern 208
   Thrush 208
Nightjar, Egyptian 144
   European 144
   Nubian 144
Noddy, Brown 114
   Common 114
   Lesser 114
Numenius arquata 98
   madagascariensis 98
   phaeopus 98
   tenuirostris 98
Numida meleagris 24
Nycticorax nycticorax 44

O

Oceanites oceanicus 28
Oceanodroma matsudairae 28
   monorhis 28
Oena capensis 138
Oenanthe albonigra 216
   chrysopygia 218
   cypriaca 214
   deserti 212
   finschii 212
   hispanica 214
   hispanica melanoleuca 214
   isabellina 212
   leucopyga 218
   lugens 216
   lugens persica 218
   lugentoides 216
   melanura 214
   monacha 218
   oenanthe 212
   picata 216
   pleschanka 214

xanthoprymna 218
Onychognathus tristramii 200
Onychoprion anaethetus 128
    fuscata 128
Oriole, Black-naped 164
    Eurasian Golden 164
    Indian Golden 164
Oriolus chinensis 164
    kundoo 164
    oriolus 164
Osprey, Western 54
Otus brucei 146
    pamelae 146
    scops 146
Ouzel, Ring 202
Owl, Arabian Scops 146
    Bruce's Scops 146
    Desert 148
    Eurasian Scops 146
    Hume's 148
    Lilith 150
    Little 150
    Long-eared 150
    Omani 148
    Pallid Scops 146
    Short-eared 150
    Striated Scops 146
    Western Barn 146
Oystercatcher, Eurasian 84

P

Painted-snipe, Greater 108
Pandion haliaetus 54
Parakeet, Alexandrine 140
    Plum-headed 140
    Ring-necked 140
    Rose-ringed 140
Partridge, Arabian 26
    Chukar 26
    Sand 26
Passer domesticus 224
    domesticus indicus 224
    hispaniolensis 224
    montanus 224
Pastor roseus 198
Pelagodroma marina 28
Pelecanus crispus 48
    onocrotalus 48
    rufescens 48
Pelican, Dalmatian 48
    Great White 48
    Pink-backed 48
Pernis apivorus 54
    ptilorhynchus 54
Petrel, Black-bellied Storm 28
    Jouanin's 32
    Matsudaira's Storm 28
    Swinhoe's Storm 28
    White-faced Storm 28
    Wilson's Storm 28

Petrochelidon fluvicola 178
Petronia, Chestnut-shouldered 224
Phaethon aethereus 48
Phalacrocorax carbo 50
    nigrogularis 50
Phalarope, Grey 110
    Red 110
    Red-necked 110
    Wilson's 110
Phalaropus fulicarius 110
    lobatus 110
    tricolor 110
Phoeniconaias minor 34
Phoenicopterus roseus 34
Phoenicurus erythronotus 208
    ochruros 208
    phoenicurus 208
    phoenicurus samamisicus 208
Phylloscopus borealis 184
    collybita 182
    collybita abientina 182
    collybita tristis 182
    fuscatus 184
    humei 184
    inornatus 184
    neglectus 182
    nitidus 184
    orientalis 182
    sibilatrix 182
    trochiloides 184
    trochilus 182
Pica pica 168
Pigeon, Bruce's Green 140
    Common Wood 136
    Yellow-bellied Green 140
Pintail, Northern 22
Pipit, Blyth's 236
    Buff-bellied 238
    Golden 236
    Long-billed 236
    Meadow 238
    Olive-backed 238
    Red-throated 238
    Richard's 236
    Tawny 236
    Tree 238
    Water 238
Platalea alba 38
    leucorodia 38
Plegadis falcinellus 38
Ploceus galbula 226
    manyar 226
Plover, American Golden 88
    Black-bellied 88
    Caspian 92
    Common Ringed 90
    Eurasian Golden 88
    European Golden 88
    Great Stone 76
    Greater Sand 92

Grey 88
Kentish 90
Lesser Sand 92
Little Ringed 90
Pacific Golden 88
Pluvialis apricaria 88
   dominica 88
   fulva 88
   squatarola 88
Pochard, Common 24
   Red-crested 24
Podiceps cristatus 34
   nigricollis 34
Porphyrio alleni 82
   madagascariensis 80
   poliocephalus 80
   porphyrio 80
Porzana fusca 78
   parva 78
   porzana 78
   pusilla 78
Pratincole, Black-winged 112
   Collared 112
   Little 112
   Oriental 112
   Small 112
Prinia gracilis 180
Prinia, Graceful 180
Prunella atrogularis 230
   ocularis 230
Psittacula cyanocephala 140
   eupatria 140
   krameri 140
Pterocles alchata 134
   coronatus 134
   exustus 134
   lichtensteinii 134
   senegallus 134
Ptyonoprogne obsoleta 176
   rupestris 176
Puffinus persicus 30
Pycnonotus cafer 166
   leucogenys 166
   leucotis 166
   xanthopygos 166

Q
Quail, Common 26
   Harlequin 26
Quelea quelea 226
Quelea, Red-billed 226

R
Rail, Water 78
Rallus aquaticus 78
Ramphocoris clotbey 170
Raven, Brown-necked 168
   Fan-tailed 168
Recurvirostra avosetta 84
Redshank, Common 100

Spotted 100
Redstart, Black 208
   Common 208
   Ehrenberg's 208
   Eversmann's 208
Remiz macronyx 166
Rhynchostruthus percivali 242
Riparia chinensis 176
   diluta 176
   paludicola 176
   riparia 176
Rissa tridactyla 116
Robin, Black Scrub 206
   European 206
   Rufous Scrub 206
   Rufous-tailed Scrub 206
   White-throated 206
Rockfinch, Pale 224
Roller, European 154
   Indian 154
   Lilac-breasted 154
   Lilac-throated 154
Rosefinch, Common 240
Rostratula benghalensis 108
Ruff 110
Rynchops albicollis 114
   flavirostris 114

S
Sanderling 104
Sandgrouse, Chestnut-bellied 134
   Crowned 134
   Lichtenstein's 134
   Pin-tailed 134
   Spotted 134
Sandpiper, Baird's 108
   Broad-billed 106
   Buff-breasted 108
   Common 102
   Curlew 106
   Green 102
   Marsh 100
   Pectoral 108
   Sharp-tailed 108
   Terek 102
   Wood 102
Sarkidiornis melanotos 16
Saxicola caprata 210
   maurus 210
   maurus armenicus 210
   maurus hemprichii 210
   maurus variegatus 210
   rubetra 210
   rubicola 210
Scolopax rusticola 94
Scotocerca inquieta 180
Serin, Yemen 240
Shearwater, Cory's 30
   Flesh-footed 32
   Persian 30

Scopoli's 30
Sooty 32
Streaked 30
Wedge-tailed 32
Shelduck, Common 20
   Ruddy 20
Shikra 64
Shoveler, Northern 22
Shrike, Bay-backed 162
   Black-crowned Bush 158
   Brown 160
   Daurian 160
   Isabelline 160
   Lesser Grey 162
   Long-tailed 162
   Masked 160
   Red-backed 160
   Red-tailed 160
   Southern Grey 162
   Steppe Grey 162
   Turkestan 160
   Woodchat 160
Silverbill, African 228
   Indian 228
Siskin, Eurasian 240
Skimmer, African 114
   Indian 114
Skua, Arctic 132
   Brown 132
   Long-tailed 132
   Pomarine 132
   South Polar 132
Skylark, Eurasian 172
   Oriental 172
   Small 172
Snipe, Common 94
   Great 94
   Jack 94
   Pin-tailed 94
Sparrow, Eurasian Tree 224
   House 224
   Indian House 224
   Pale Rock 224
   Spanish 224
   Yellow-throated 224
Sparrow-Lark, Black-crowned 172
Sparrowhawk, Eurasian 64
Spilopelia senegalensis 138
Spinus spinus 240
Spoonbill, African 38
   Eurasian 38
Starling, Amethyst 198
   Brahminy 200
   Chestnut-tailed 200
   Common 198
   Rose-coloured 198
   Rosy 198
   Superb 198
   Tristram's 200
   Violet-backed 198

   Wattled 198
Stercorarius antarcticus 132
   longicaudus 132
   maccormicki 132
   parasiticus 132
   pomarinus 132
Sterna dougallii 126
   hirundo 126
   paradisaea 126
   repressa 126
Sternula albifrons 128
   saundersi 128
Stilt, Black-winged 84
Stint, Little 104
   Long-toed 104
   Temminck's 104
Stone-curlew, Eurasian 76
   Great 76
Stonechat, Caspian 210
   European 210
   Pied 210
   Siberian 210
Stork, Abdim's 36
   Black 36
   White 36
   Yellow-billed 36
Streptopelia decaocto 138
   orientalis 136
   risoria 138
   roseogrisea 138
   tranquebarica 138
   turtur 136
Strix butleri 148
   hadorami 148
Sturnia malabarica 200
   pagodarum 200
Sturnus vulgaris 198
Sula dactylatra 50
   leucogaster 50
   sula 50
Sunbird, Nile Valley 222
   Palestine 222
   Purple 222
   Shining 222
Swallow, Barn 178
   Indian Cliff 178
   Lesser Striped 178
   Red-rumped 178
   Streak-throated 178
   Wire-tailed 178
Swamphen, African 80
   Grey-headed 80
   Purple 80
Swan, Bewick's 16
   Mute 16
   Whooper 16
Swift, Alpine 152
   Common 152
   Eastern Common 152
   Forbes-Watson's 152

Fork-tailed 152
Little 152
Pacific 152
Pallid 152
Sylvia althaea 194
  atricapilla 192
  borin 192
  communis 194
  crassirostris 194
  curruca 194
  curruca halimodendri 194
  leucomelaena 194
  melanocephala 196
  minula 194
  mystacea 196
  nana 192
  nisoria 192

T

Tachybaptus ruficollis 34
Tachymarptis melba 152
Tadorna ferruginea 20
  tadorna 20
Tchagra senegala 158
Tchagra, Black-crowned 158
Teal, Cotton 18
  Eurasian 22
Tern
  Arctic 126
  Black 130
  Bridled 128
  Caspian 124
  Common 126
  Greater Crested 124
  Gull-billed 124
  Lesser Crested 124
  Little 128
  Roseate 126
  Sandwich 124
  Saunders's 128
  Sooty 128
  Swift 124
  Whiskered 130
  White-cheeked 126
  White-winged 130
  White-winged Black 130
Terpsiphone viridis 166
Tetrax tetrax 76
Thalasseus bengalensis 124
  bergii 124
  sandvicensis 124
Thick-knee, Spotted 76
Threskiornis aethiopicus 38
Thrush, Black-throated 202
  Blue Rock 204
  Common Rock 204
  Dark-throated 202
  Dusky 202
  Eyebrowed 202
  Mistle 204

Red-throated 202
Rufous-tailed Rock 204
Song 204
White's 204
Tit, Black-headed Penduline 166
Tmetothylacus tenellus 236
Todiramphus chloris 156
Torgos tracheliotos 52
Treron waalia 140
Tringa erythropus 100
  flavipes 100
  glareola 102
  nebularia 100
  ochropus 102
  stagnatilis 100
  totanus 100
Tropicbird, Red-billed 48
Turdoides squamiceps 164
Turdus atrogularis 202
  eunomus 202
  obscurus 202
  philomelos 204
  ruficollis 202
  torquatus 202
  viscivorus 204
Turnix sylvaticus 24
Turnstone, Ruddy 102
Tyto alba 146

U

Upupa epops 154

V

Vanellus cinereus 86
  gregarius 86
  indicus 86
  leucurus 86
  spinosus 86
  vanellus 86
Vulture, Cinereous 52
  Egyptian 52
  Eurasian Black 52
  Griffon 52
  Lappet-faced 52
Wagtail, Amur 232
  Black-backed Citrine 230
  Black-headed 234
  Citrine 230
  Eastern Yellow 232
  Forest 230
  Grean-headed 232
  Grey 230
  Grey-headed 234
  Masked 232
  Sykes's 234
  Western Yellow 234
  White 232
  White-headed 234
  Yellow-headed 234
Warbler, Arabian 194

Arctic 184
Asian Desert 192
Barred 192
Blyth's Reed 188
Booted 190
Caspian Reed 188
Cetti's 180
Clamorous Reed 186
Common Grasshopper 188
Dusky 184
Eastern Bonelli's 182
Eastern Olivaceous 190
Eastern Orphean 194
Eurasian Reed 188
Fan-tailed 180
Garden 192
Great Reed 186
Green 184
Greenish 184
Hume's Leaf 184
Icterine 192
Indian Reed 186
Marsh 188
Ménétriés's 196
Moustached 186
Olive-tree 190
Paddyfield 186
Plain Leaf 182
Red Sea 194
River 188
Sardinian 196
Savi's 188
Scrub 180
Sedge 186
Streaked Scrub 180
Sykes's 190
Thick-billed 190
Upcher's 190
Willow 182
Wood 182
Yellow-browed 184
Watercock 80
Waterhen, White-breasted 80
Waxbill, Common 228
Weaver, Ruppell's 226
Streaked 226

Wheatear, Arabian 216
Black-eared 214
Cyprus 214
Desert 212
Eastern Black-eared 214
Eastern Mourning 218
Eastern Pied 214
Finsch's 212
Hooded 218
Hume's 216
Isabelline 212
Kurdish 218
Kurdistan 218
Mourning 216
Northern 212
Pied 214
Red-tailed 218
Rufous-tailed 218
South Arabian 216
Variable 216
White-crowned 218
Whimbrel 98
Whinchat 210
White-eye, Abyssinian 196
Oriental 196
White-breasted 196
Whitethroat, Common 194
Desert 194
Hume's 194
Lesser 194
Wigeon, Eurasian 20
Woodcock, Eurasian 94
Wryneck, Eurasian 154

X
Xema sabini 116
Xenus cinereus 102

Y
Yellowlegs, Lesser 100

Z
Zoothera aurea 202
Zosterops abyssinicus 196
  palpebrosus 196